Praise for *The Future of Everything*

"With wit, humor, and clarity *The Future of Everything* gives a good overview of the history of science as a predictive tool. . . . Orrell's work has the potential to change the way we plan for the future, both personally and as a society." —*Winnipeg Free Press*

"Even big-picture predictions—global warming, impending pandemics, the global impact of consumerism, emerging technologies such as genetic engineering and nanotechnology—seem less overwhelming after this enlightening history lesson. . . . Some of Orrell's most eloquent and provocative passages are about those unpredictable forces known as free will and personal choice, and he rallies us all to effect positive change." —*Canadian Geographic*

"If you loathe uncertainty, can't cope with complexity, or prefer easy (yet wrong) solutions to complex problems, you may not appreciate this book. On the other hand, you will likely find it reassuring, even empowering perhaps, to know that the 'experts' can be wrong, and that common sense is still useful." —*Edmonton Journal*

"The special and enchanted land of forecasting is inhabited by tribes of specialists who speak a language that is unintelligible to most of us. David Orrell's *The Future of Everything* is a wonderful guide and companion to that land; its history, its valuable resources and its fault lines. And it still manages to be a good read." —*Everett Herald*

"Orrell's writing is top-notch, but he's at his finest when writing about the old days, somehow finding exactly the right mix of anecdote, broad brush, and humor. There's the goat's-beard hygrometer (curly plant = humid weather) and the invention of the word 'forecast' (to distance weather prediction from astrological prediction, though the same folks often did both)."   —*The Globe and Mail*

"Mathematician David Orrell explains why the mathematical models scientists use to predict the weather, the climate, and the economy are not getting any better, just more refined in their uncertainty . . . Dr. Orrell is no climate-change denier. He calls himself green. But he understands the unjustified faith that arises from the psychological need to make predictions."

—*The National Post*

THE FUTURE OF
EVERYTHING

THE FUTURE OF
EVERYTHING
THE SCIENCE OF PREDICTION

FROM WEALTH AND WEATHER
TO CHAOS AND COMPLEXITY

DAVID ORRELL, PhD

BASIC BOOKS
A MEMBER OF THE PERSEUS BOOKS GROUP
NEW YORK

The Future of Everything
The Science of Prediction
From Wealth and Weather to Chaos and Complexity

Hardcover edition published in January 2007 in the United States by Thunder's Mouth Press, by arrangement with HarperCollins Publishers Ltd., Toronto, Canada

Paperback edition published in December 2007 in the United States by Basic Books, A Member of the Perseus Books Group, 387 Park Avenue South, New York, NY 10016-8810

Library of Congress Cataloguing-in-Publication Data is available.

Hardcover: ISBN-10: 1-56025-975-2; ISBN-13: 978-1-56025-975-6
Paperback: ISBN-13: 978-1-56858-369-3
9 8 7 6 5 4 3 2 1

Printed in the United States of America

This book is dedicated to the memory of my father,
John Orrell, to Beatriz and to Isabel

Contents

FUTURE

► INTRODUCTION
THE SCIENCE AND SOCIOLOGY OF FORECASTING

The term "natural disaster" has become an increasingly anachronistic misnomer. In reality, human behavior transforms natural hazards into what should really be called unnatural disasters.
—Kofi Annan, Secretary-General of the United Nations

Prediction is very difficult, especially if it's about the future.
—Niels Bohr, Nobel laureate in physics

ANATOMY OF A STORM

By December 15, 1999, the run-up to the new millennium had begun in earnest. People around the world were getting the champagne in, ready to uncork it the moment 1/1/2000 did its giant, slow march into their time zone. The NASDAQ stock-market index was also acquiring a champagne-like froth, as the Internet rewrote the rules of the world economy. The only cloud on the horizon—a potential storm—was the Millennium Bug, a software error caused when programmers rounded off the computer's internal date from four digits to two. Many predicted it would bring chaos, or even collapse, to the world economy.

In the slums, or *ranchos,* perched high on El Avila mountain north of Caracas, Venezuela, the Millennium Bug was not a major worry. Most of the residents had electricity, acquired by tapping illegally into the network, but Internet access was limited, to say the least. They were more excited about that day's referendum, which their hero, the new president, Hugo Chavez, had called to get the constitution approved. Voter turnout was high, and 78 percent in favour, despite the heavy rains, which had been falling for days and seemed to be getting worse.

Normally, the rainy season lasted only until October, but this year was an exception. The weather was out of synch. Some in the government suggested that the more vulnerable areas of Caracas be evacuated, in case the steep slopes became unstable. Perhaps not wishing to disrupt the referendum, however, the government took no action.

Early the next morning, December 16, the northern side of El Avila, which faces the coastal resorts near the airport, simply gave way. Witnesses said that huge waves of water, six metres high, cascaded down the mountain, carrying away everything in their wake—trees, cars, houses, people. Giant boulders hurtled down narrow gullies, not stopping until they fell into the ocean or smashed into the luxury apartment blocks that lined the coast. One survivor, a young woman, described waking up in the middle of the night to the sound of the water and the rocks crashing down the mountain and people yelling, "The river is coming!"[1] The water filled half her house before she could get out the door. Many others weren't so lucky.

At first, the scale of the disaster was not comprehended. Estimates were for 100 dead, then 500. By December 22, this had grown to 30,000, with perhaps ten times that many left homeless. Many of the bodies were swept out to sea or buried under the mud, and were impossible to recover. Survivors gathered in halls and stadiums, desperately searching for lost relatives. Rescue workers were completely overwhelmed. One told a journalist from the *Independent* how he had rescued a three-year-old girl: "Every time she saw water, she screamed."[2] Many feared that the lack of drinking water and sanitation would lead to disease outbreaks, magnifying the human impact, but fortunately this did not come to pass.

Almost immediately, the storm was politicized. The president's opponents castigated him for proceeding with the referendum. On Christmas Day, Chavez distributed gifts to hundreds of orphaned children at the Poliedro sports arena, in Caracas. When a reporter suggested the ex-paratrooper was culpable for the disaster, he replied: "They should shoot me if I have any personal responsibility in this."[3]

Could this storm, with all of its social, economic, and medical repercussions, have been predicted? Was it a random, unforeseeable event, or was someone responsible? How about the weather forecasters? The conditions for the storm began to develop in early December, when a cold front encountered a southwesterly flow of moist air, resulting in precipitation over the northern coast. There was one week of moderate rain, followed by two days of extremely heavy rain on December 15 and 16. The daily totals recorded at the nearby Maiquetia International Airport for these days were so excessive that, in theory, they wouldn't be repeated for 1,000 years—truly a millennial storm.[4] Almost by definition, such events do not get predicted; no forecaster likes to predict something he has never seen happen.

High above the storm, watching it develop, was a GOES 8 satellite belonging to the National Oceanic and Atmospheric Administration. Its mechanical infrared eye didn't pick up the horror on the ground; it was focused on the cloud-tops, taking temperature readings that could be used to estimate rainfall. However, the relatively coarse resolution—the smallest features it could detect were roughly four kilometres by four kilometres—meant that it could only confirm the location of the maximum rainfall.[5] The storm seemed to stand still, as if it was intent on bringing the mountain down.

Even with the amount of rain, though, no one could have foreseen the scale of the mudslides. And if they had, no one would have paid attention. As a forecaster at the National Hurricane Center in Miami admitted, for every mudslide they got right, they were "going to end up screaming wolf maybe 10 times."[6] The instability of the soil was the result of a number of factors, not all of them natural. The poor *rancho* areas were completely unregulated by the government; in fact, because of high levels of crime, many were no-go zones for the police.[7] Much of the forested land around the hills had been clear-cut by residents for firewood and construction material or cut down by private companies, weakening the soil. Benches had been cut into the steep hillsides to support the houses of cinder block and corrugated iron, further destabilizing the ground. Many of the homes were situated in dry riverbeds, so were directly in the path of the mudslide as it coursed down the gullies.

In fact, the disaster was caused by a range of complex, intertwined forces. Historical records from Spanish archives show that major floods and landslides are hardly new to the area.[8] Casualties were so high this time because of the sheer number of people—around 3 million—who now make their homes on the highly inhospitable mountains clustered around Caracas. This is a result not of extreme weather but of extreme disparities in wealth and migration to overpopulated urban areas. The storm, and its unusual timing, might also have been influenced by global warming, which in turn depends on the amount of carbon dioxide emitted from the world economy. Warmer oceans mean more water gets evaporated, so storms are expected to become more powerful.[9] The stakes for forecasters may grow higher still.

When I visited the flood area a year later, it was still covered in a layer of light brown soil. Since it was impossible to recover bodies, some of which were buried six metres down or more, the Catholic priests consecrated the entire place as a burial site. Already, though, the *ranchos* are encroaching on the territory they lost. Was it really a once-in-a-millennium event, or will the same thing happen again in the next hundred years? Or the next ten?

MAPPING THE FUTURE

The Future of Everything is about scientific prediction in the areas of weather, health, and wealth—how we foresee storms or fair weather, sickness or health, booms or crashes. It might seem that forecasts of the atmosphere have little to do with prediction of diseases or the economy, but in fact these three areas are closely linked. For one thing, they often affect each other, so prediction is an intrinsically holistic business. As shown above, a storm's impact depends on the conditions on the ground, and can have huge economic consequences. When Hurricane Katrina swung into the Gulf coast in late August 2005, flooding much of New Orleans and knocking out oil refineries, its financial impact was greater than that of the 9/11 terrorist attacks. In 2003, for Toronto, Hong Kong, and other cities, the storm was called SARS. Global warming too is a multi-stranded problem with complex repercussions. Like a potentially larger version of the Caracas storm, it exists at the centre of a vortex of social and environmental causes and effects. Atmospheric carbon dioxide is influenced by economic output and population levels; the resulting climate shifts and environmental stress may affect the spread of disease; large-scale epidemics have in the past severely disrupted economic activity; and so on.

The three types of prediction also use similar methods and share a common past. A traditional technique is astrology, which links the biological event of a baby's birth or the atmospheric and economic event of good harvest weather to the motion of the planets. Many people begin their day by reading their horoscope in the newspaper. Because I'm a Gemini (and therefore conflicted), I read my horoscope occasionally but never believe it. However, I will gladly check the five-day, long-range weather forecast, which probably has a lower accuracy rate.

For much of history, the same experts supplied both horoscopes and weather forecasts; humans and the atmosphere danced to the same tune. The seventeenth-century astronomer Johannes Kepler paid his way through university, and indeed much of the rest of his life, this way. Even now, when we talk about the weather, we often describe it in almost human terms. It is a cheerful day or a gloomy one; a storm is violent; a hurricane has a mind—and always a name—of its own. The weather is a character in all our lives, and sometimes it is a criminal.

The first newspaper weather map appeared on April 1, 1875, in *The Times* of London. It was prepared by the British scientist Sir Francis Galton, who also discovered the anti-cyclone.[10] Three years later, Galton gave a lecture at London's Anthropological Institute on a rather different subject. He presented composite photographs of the faces of prison convicts, dividing them into three groups by the type of offence. His aim was to search for common characteristics within each group. Influenced and inspired by the ideas of his cousin, Charles Darwin, Galton believed that traits such as "eminence" and criminality were inherited and linked to physical appearance. Just as he had scanned weather maps for patterns that would foretell a coming storm, he now looked for facial characteristics that warned of criminality. To help bring out such features, he made the composite pictures, which overlaid the faces of up to eight individuals.[11] The results were not very useful for prediction. What they showed was more a common humanity than any particular demonic trait. (As discussed in Chapter 5, this didn't prevent Galton from inventing the field of eugenics to "improve" the human race.)

Since Galton's time, a huge scientific effort has been devoted to looking into our future weather, health, and wealth, now using mathematical models. These emulate the flow of air and water in the atmosphere, or substances in our body, or money in the economy, using large sets of equations. Although the calculations are performed on high-speed computers, the techniques are essentially the same as those first developed by physicists such as Isaac Newton to study the dynamics of celestial objects. Like astrology, our predictive models of the future have their roots in the stars.

Weather prediction has evolved in the past half century into the multi-billion-dollar business of providing up-to-the-minute forecasts to the media and to weather-affected industries, such as agriculture, transport, and insurance. In biology, the Human Genome Project catalogued all human genes in a kind of giant library of our species; one of the stated aims was the prediction and control of genetic traits and diseases. Perhaps the greatest preoccupation of predictors has been the infinitely intriguing motions of the trillion-dollar financial markets. Companies, governments, and universities around the world, as well as giant institutions like the International Monetary Fund and the World Bank, hire thousands of economists in an effort to foresee economic events.

It turns out, though, that predictions of weather, health, and wealth have another thing in common. While scientists have had great success in squinting through microscopes at the smallest forms of life, or smashing atoms together in giant particle accelerators to analyze the structure of matter, or using telescopes to look forwards in space and backwards in time at the formation of distant galaxies, their visions into the future have been, like Galton's composite photographs, blurred and murky. As a result, projections tend to go astray.

In weather forecasting, for example, accuracy has improved in a slow, iterative manner, but if you're not a natural risk-taker, you should put little faith in a five-day forecast. We can put a man on the moon, but timing a shuttle landing around the weather is still tricky. In medicine, biologists have realized that the connection between genes and traits is not a straight line but a highly twisted and circuitous one. It is frequently announced in the press that the gene that causes some condition has been discovered, only for the news to fade from public attention as the complexities emerge. And in economics, the dominant "efficient market" theory, which has been weirdly embraced by many highly paid predictors, says that, in principle, the economy cannot be predicted. The best one can do is predict and control financial risk. Even this aim seems out of reach after events like the Black Monday crash in 1987 or the bursting of the Internet bubble.

The reason scientists—who are usually not closet Nostradamuses—have been drawn to making predictions in these areas (rather than, say, fashion or popular music) is because the underlying systems seem quantifiable and computationally tractable. Weather is just fluid flow, the human body is biochemistry, the economy is money. So what is going wrong? What do these systems have that escapes the models? Is our difficulty in forecasting the future health of the planet related to our difficulty in predicting the health of our own body? Or the health of the economy? Will we always be blind to the future, reacting impulsively to the next crisis or piece of good fortune when it comes along? Finding the answers to these questions is the target of *The Future of Everything*.

Don't Blame the Butterfly

My own introduction to predictability occurred when I returned to university, after several years as a jobbing mathematician, to do a Ph.D. on model error in weather forecasting. Model error represents the difference between the model—typically a set of mathematical equations based on physical "laws"—and the actual system it is supposed to emulate. For example, the trajectory of an arrow is something that can be determined reasonably accurately from the arrow's starting position and its velocity—the initial conditions—using the laws of physics. But if there's a gust of wind that is not included in the model, then the arrow will depart slightly from its predicted path. That's model error. It might not be important, unless you happen to be the person waiting at the other end with the apple on your head.

In weather forecasting, there had been little investigation of model error, despite the fact that even forecasters agreed that predictions usually missed their target after about two or three days. From my work experience (by which I mean glaring discrepancies between my calculations and reality), I knew that even engineered systems, where all the forces and material properties are exactly known, could still be hard to model accurately. The atmosphere was a horrendously complex system in comparison, so model error should have been huge. However, the dominant theory was that forecasts went wrong not because of any deficiency in the model, but because small errors in the initial condition were magnified by chaos—the so-called butterfly effect. Storms like the one that hit the north coast of Venezuela could, in principle, be caused by an insect flapping its wings somewhere on the other side of the world.

My work over the next couple of years was aimed at developing a technique for measuring model error that filtered out the effects of chaos. When our group's results (which showed that most forecast error was a result of the model, with chaos a relatively minor effect) were accepted for publication and presented at conferences, there was initially no reaction from the meteorological community. But then the story was picked up by the media. Soon there were reports in newspapers and magazines and on radio shows in Europe, North America, and elsewhere. The reaction seemed out of proportion to the actual scientific interest; but everyone is interested in the weather, and everyone knows that the forecasts are wrong. The idea that the cause could be the models was apparently big news. Perhaps storms could be better predicted.

While this seemed a positive development, it didn't go down well. Criticizing models was apparently a good way to annoy the weather gods, or at least the weathermen.[12] Eventually, over a few years, the storm clouds dissipated. The work was published, and I moved on to different things. However, I remained struck by the deep blanket of denial that settled over those in the meteorological community, and by their emotional reaction to criticism, where any questioning of the model was interpreted as a personal attack. No matter the evidence to the contrary, they always believed that their models were right. It raised questions in my mind about the science and sociology of forecasting, questions this book attempts to answer.

I learned later that this lack of zeal in investigating model error was not unique. The work of Will Keepin and Brian Wynne at the International Institute for Applied Systems Analysis (IIASA) in Austria was proof of that. In 1981, the institute had just finished a multi-million-dollar project, involving over a hundred scientists, that was supposed to forecast the world's future energy consumption. After accounting for factors such as demographics and projected oil reserves, the computer model predicted that demand for energy would rise enormously over the next fifty years, and could be met only by building over a hundred nuclear power plants a year. Needless to say, this would have required a huge expansion in the nuclear industry, creating lucrative jobs for anyone with expertise in the energy field—such as, for example, the scientists who had built the model.

To Keepin, however, the model was so flexible, its connection to reality so tenuous, that "it was a bit like the Wizard of Oz. . . . Some guy was pulling on levers and making a big show, but it was a show determined by the little guy behind the curtain."[13] Almost alone in his criticism of the model, he decided to resign from his job. Just then, the British scientist Brian Wynne came to IIASA on a two-year contract to study the politics of science. He was looking for an insider to tell all about the energy model. After hearing Keepin's story, Wynne decided, to the institute's horror, to devote his two years to studying the other scientists' reaction to Keepin's critique. In 1984, the results were published in the top journal *Nature*.[14] The paper showed that the model had been biased in favour of nuclear and fossil-fuel energy producers, and that the model developers had tried to conceal its shortcomings.

Similarly, when the mathematician Benoit Mandelbrot questioned the assumptions behind modern finance, he found himself "about as welcome in the established church of economics as a heretical Arian at the Council of Nicene."[15] As the philosopher Thomas Kuhn pointed out, in science it is common, and even healthy, for new ideas to be met with skepticism by the establishment.[16] But if scientific models are used to set policy, and to make important public and private decisions, then we need to know how accurate they are. This is made difficult by the nature of the models, which are written in a highly specialized language that can be understood only by other scientists with experience in the field. Since such people often share the biases of the model makers, glaring problems go undetected, or unremarked.[17] Indeed, scientific institutions have become expert at deflecting serious public scrutiny of their work.[18] So to what extent can we trust their predictions for the future?

WHERE IS THIS GOING?

Forecasting has always attracted fraudsters and con men. When Kepler was trying to promote his predictive model of the sun, moon, and planets to Emperor Rudolph II, his competition was not so much other astronomers, but savants like the Englishman Edward Kelley, who preferred a talking mirror to calculations and was eventually jailed by the emperor for his poor performance. More recently, studies have shown that social forecasting, scientific and otherwise, is about as accurate as random guessing, despite the vast numbers of highly paid experts employed to do it.[19] If the futurologists of the 1960s had been right, for example, I would probably be writing this in an orbital space station as my personal robot tended to my toenails.

The accuracy of forecasts would not be so important if all that were at stake was the weekend weather or the likely return from government bonds in the next quarter. But like the residents of Caracas, we are becoming aware that the future need not resemble the recent past; the coming storms in weather, health, and wealth may be more intense than the kind we have grown used to.

It is only in the past few decades that human activities, like our use of large-scale industry and the automobile, have become comparable in scale to the workings of the planet itself. And it is even more recently that we have started to learn of holes in the ozone layer, the spread of chemical pollutants through the food chain, and the collapse of ocean fisheries because of overfishing. We have passed a kind of tipping point in our relationship with the world; our actions now influence its workings at every level. It used to be that the world happened to us; now we happen to the world as well. One day, as our children survey the damaged planet they have inherited, we may hear the question (asked less fairly of the Venezuelan president), Are you responsible?

The next fifty or one hundred years are going to be crucial, and we need to have a guide. In many ways, science got us into this fix, but will it help us to get out? And even if it can't tell us exactly where the world is headed, can it help us predict our own future health or play the stock market? To answer these questions, it is necessary to understand how scientists go about making forecasts. Equally important, though, is the history and sociology of science. Like any complex process, prediction is path-dependent: it matters how we got here.

My own short experience with the weathermen was, in the scale of things, a minor affair, a tempest in a teacup. I wasn't burned at the stake like Giordano Bruno, who had tried to convince the Inquisition that space was infinite, or threatened with torture and imprisonment for mocking the pope and arguing that the earth went around the sun, like Galileo Galilei. However, it did make me realize that in many ways, science has become rather like the Catholic Church of Galileo's time, and about as receptive to criticism. And just as we once looked to the Church to predict the future—just keep your head down until the Second Coming—we now look to the scientists for guidance.

We are all predictors, living by our forecasts. The most primitive bacteria have the ability to sense the presence of food and move towards it. Living beings are constantly interacting with their environment, reading and displaying information. Successful strategies—knowing when to hunt, when to run, when to sit it out—are coded in the genes. The practice of speed dating is based on the idea that first impressions count: within a few minutes of meeting a prospective mate, we somehow fast-forward through the whole relationship and predict whether it will work (and because this affects how much effort we put into the relationship, it can become a self-fulfilling prophecy).[20] We may not judge a book by its cover, but it certainly helps to scan the first few pages for a summary.

In science, though, forecasting plays a special role. Predicting the future is not a side activity of science, but has come to be seen as its primary pursuit.[21] A scientific theory is generally considered valid only if it can be used to predict the behaviour of a system. A theory that doesn't predict may be a beautiful or elegant idea, but it's no more functional, in the view of many scientists, than a piece of modern art. We may all be predictors, but for scientists, it's their profession. In this book, however, I will argue the following:

- **Mathematical models interpret the world in simple mechanical terms.** Scientific prediction, from ancient astronomy up to and including chaos theory, has been based on a highly abstracted, mechanistic view of the world, which is of limited applicability in the context of complex systems.
- **Living things have properties that elude prediction.** Systems where predictions are of interest—in biology, economics, or climate change—are either alive, influenced by life, or have a similar level of complexity to living beings. They are difficult to predict not because of simple technical reasons, which can be overcome with faster computers or better data, but because they have evolved to be that way. We pinpoint the causes of prediction error.
- **Forecasting has a large psychological component.** The desire to explain the world in terms of simple cause-and-effect relationships is a fundamental characteristic of human beings.

Predictions often tell us more about group psychology than they do about reality. Many prognosticators anticipated chaos in the financial markets in the new millennium, but the cause was supposed to be the Millennium Bug, not the collapse of Internet stocks. And accurate predictions, such as those that pointed out the vulnerability of New Orleans to a hurricane, are often ignored.[22]

- **Some predictions are still possible.** One type of prediction relates to overall function and can be used to make general warnings. The other type involves specific forecasts about the future. Mathematical models are better at the first than they are the second (Niels Bohr was right: predicting the future is hard).

- **We need to change our approach to prediction.** The current debate between climate modellers who argue that global warming is an imminent threat and skeptics who demand further proof can be resolved only with a fundamental shift in the kinds of predictions we make.

This book is divided into three main parts. The first is a brief history of the science of prediction. It will argue that modern forecasters are drawing on a long tradition of modelling the physical universe that stretches back to the ancient Greeks; and that throughout history forecasters have not just peered into the future but have helped shape the world we live in. Everything from our economic system to our relationship with nature and our own bodies has been profoundly affected by the early predictors, the model makers, the champions of cause and effect.

Of course, reading any such abbreviated history is a little like listening to a classic rock station on the radio. Just as each band is allowed to have only a handful of representative songs, so the great scientists have their life's work boiled down to a couple of greatest hits: Pythagoras and the Music of the Spheres, Kepler and his Harmony of the World, Galileo and his Stones. Minor scientists—the supporting acts—seem never to have existed, except as part of the occasional quirky sideshow. Unlike most classic rockers, though, the great scientists considered here are all European males. This is not because prediction is not practised by other races, or indeed by females, but because culture has played a role in the development of prediction as it is currently practised by scientists. And as the science historian Evelyn Fox Keller has pointed out, science has not been a gender-neutral pursuit.[23] These are subjects we will return to.

The second part of the book examines forecasting practice in the specific areas of weather, health, and wealth, and describes in detail the techniques currently employed by the scientists who make prediction their living. Like siblings, these three main areas of scientific prediction grew up together, share DNA, and show similar traits. To understand one, it helps to know the others. Finally, in the third part of the book, we see how these separate strands come together in a long-term forecast for the planet—culminating in a look at predictions for the year 2100.

The ultimate aim of the book is to make a forecast about forecasting, and to try to answer the question, Can scientists really look into the future? To find the answer, we must begin with the spiritual and intellectual forebears of modern numerical prediction—a secretive cult in ancient Greece led by a man they claimed was the son of Apollo.

PAST

1 ► SLINGS AND ARROWS
THE BEGINNINGS OF PREDICTION

All things are full of gods.
 —Thales, Greek philosopher and mathematician

The truth of the model is not the truth of the phenomenon. It is a common confusion between these two kinds of truth—the norm in magic—that sometimes sanctifies the model (which is regarded as part of the real world) and gives the scientist the role of priest.
 —Antoine Danchin, Pasteur Institute biologist

GAIA

According to Greek mythology, the first oracle, the maker of forecasts, was the earth goddess Gaia. She held forth at Delphi, which was named after the Greek word *delphus,* for "womb," and was literally the womb of the earth. Geographically, Delphi is located on a gentle slope on Mount Parnassus, about 150 kilometres northwest of Athens. On one side, the area is towered over by 300-metre cliffs that are known as the Phaedriades, or Shining Ones, because of the almost metallic way they catch the morning and evening light. The ground is nourished by the Castalian spring, which flows through a cleft in the cliffs. Below, a gorge filled with olive trees leads down to the Gulf of Corinth. The whole area is prone to storms, landslides, and other outbursts of the gods. It is watched over by birds of prey who ride the thermals of the cliffs.

The ancient Greeks believed this beautiful and dramatic place to be the centre of the universe. A legend states that the god Zeus released two eagles, one from the east, one from the west. When they met at Delphi, Zeus placed a stone, the *omphalos,* to mark the spot. Gaia's prophecies were sung out by a mythical figure referred to as Sybil, who inhaled trance-inducing vapours from a fissure in the mountain. The site was guarded by Gaia's daughter, the fearsome serpent Python, who lurked in the nearby Castalian spring.

Like his father Zeus, the god Apollo had an interesting and complicated life. He was god, among other things, of reason, music, plague, and archery. He had many love affairs, with both goddesses and mortal humans. But the young, inexperienced god's first big achievement—the one that put him on the map—was to slay the giant serpent Python:

> E're now the God his arrows had not try'd
> But on the trembling deer, or mountain goat;
> At this new quarry he prepares to shoot.
> Though ev'ry shaft took place, he spent the store
> Of his full quiver, and 'twas long before
> Th' expiring serpent wallow'd in his gore.[1]

Since Python was Gaia's daughter, amends had to be made for this violent deed. Apollo worked for eight years as a cowherd to purify himself. But once that was done, he returned to Delphi and, in a hostile takeover, claimed the oracle from Gaia. From that moment on, he was known as Pythian Apollo, the god of prophecy, and Delphi was his main shrine.[2]

That's the mythology. Archaeological excavations have shown that from 1500 to 1100 B.C., the site was occupied by small Bronze Age Mycenean settlements that were dedicated to the Mother Earth deity. The new god Apollo arrived, perhaps via invading Dorians, and began to dominate. So in both versions, a power shift takes place between Gaia and Apollo. The chaos theorist Ralph Abraham refers to this time in human history as a major bifurcation point, where "the goddess submerged into the collective unconscious, while her statues underwent gender-change operations."[3] The result was the most successful prediction business in history. For almost a thousand years, the Delphic Oracle called the shots in business, politics, religion, and war.

The biographer Plutarch, best known for his lives of famous Greeks and Romans, served as a priest at Delphi, and from his histories we have some knowledge of the inner workings of the Delphic sanctuary.[4] The oracle, known as the Pythia, was always a woman, since women were thought to be more receptive to Apollo's oracular powers. Like a telegenic TV presenter, the Pythia didn't make the forecasts herself, but only channelled the predictive power of Apollo. The main job requirements were *enthousiasmos* (which in its original sense meant not enthusiasm but "possessed by a god") and faithfulness to Apollo. She was not allowed to have intimate relations with anyone, even a husband, for Apollo was a jealous god. A case in point was Cassandra. Apollo attempted to seduce her by granting her prophetic powers, but she refused him. In revenge, he cursed her so that no one would pay attention to her predictions.

The oracular ceremonies were held once per month, except during the three-month winter break, when Delphi was often covered in snow. Suppose you are a *theoprope,* a supplicant. You arrive by boat at the harbour of Kirrha, in the Gulf of Corinth, then make the journey up into the mountains, reaching Delphi as night falls.

You have with you two things: a written question and, for reasons that will become obvious, a young goat (which you purchased from a goatherd outside the town). You spend the night at a crowded inn, then get up early the next morning to join the long line of people outside the temple. In your mind is the question you have carried all this way. Perhaps it relates to a marriage, or treatment of an illness, or a business concern.

Your growing anxiety isn't helped when you notice that some people appear to be jumping the queue, after offering the priests extravagant bribes. But finally it is your turn. A priest beckons you to climb the steps of the temple. In your arms is the small, warm goat. You feel it trembling with fear. You hand it to the priest, who takes it towards a blood-stained altar. Another priest has at the ready a long bronze blade. On the walls, you notice, are inscribed rather bland motivational messages. Know thyself. Avoid excess. A single letter *E*. What can that mean? While the first two priests busy themselves with the poor struggling animal, another leads you to the spring near the temple. You have to shower before they will let you into the pool. As you wash, you try to close your ears to the goat's plaintive bleats, which are soon followed by silence. You hope that Apollo is satisfied by the humble sacrifice.

Once purified, you are led by a high priest to the inner sanctum. And there she is: the Pythia, the oracle. She sits on a three-legged bronze stool, the tripod. The room, you notice, has a peculiar sweetish smell—some strange vapour that seems to be emanating from the earth itself.[5] The Pythia is a middle-aged woman. Her hair is thin and grey, her eyes appear glazed. She doesn't seem to notice you come in. Suddenly, you feel very afraid of this person.

Before you can back out of the room, the high priest reads your question aloud. Again, the Pythia fails to react. She sways slowly back and forth on her tripod. You wonder if she has heard.

But then she starts to make a noise. Not exactly speech or singing, but something in between, on the edge of sense and nonsense. You listen, but it is like trying to make sense of the call of birds or the rustling of leaves in a storm.

After some time, you're not sure how long, the Pythia falls silent. It is as if a switch in her head just turned off. You notice how drained she looks. The high priest steps forward. Whatever language she was speaking, he must understand it, because he reads out a neat response in hexameter verse. You're trying to figure out what it means as they lead you down the steps of the temple. And you're still trying to figure it out days later, when you eventually get home. But when you announce your decision to your waiting family, it feels like you knew it all along.

According to the philosopher Heraclitus, the Pythia never gave a straight answer, but only hinted at the truth. King Croesus of Lydia famously asked the Pythia if he should invade Persian territory. The oracle told him that if he did, a mighty empire would be destroyed. He took this as a green light, but unfortunately, the empire she was referring to was his own.[6]

Despite the equivocal nature of the prophecies, the oracle played an enormously important role in Greek culture, especially in the Archaic period (the eighth to sixth centuries B.C.). Most major decisions about war or politics were made in consultation with it. The oracle retained its power for almost a thousand years, gradually falling into decline with the rise of Christianity, and in the third century A.D., it made its final prediction: the gods would no longer speak at Delphi.

APOLLO'S ARROW

The poet Iamblichus relates a tale about the oracle when it was still at the height of its powers. A gem engraver called Mnesarchus visits to ask whether a journey he is about to undertake will be profitable. The oracle replies that it will; furthermore, the man is told, his wife, who unknown to him is pregnant at the time, will give birth to a son "surpassing in beauty and wisdom all that had ever lived."[7] Mnesarchus realizes that the child has been sent by the gods. When he is born, he is named Pythagoras, "signifying that such an offspring had been predicted by the Pythian Apollo."[8]

Mathematician, philosopher, even Olympic trainer, Pythagoras would go on to found a new system of prediction based not on oracles but on the power of numbers. He was literally a demigod to the Greeks—some said he had been fathered by Apollo.[9] This was a story that his many followers never denied. A proof of Pythagoras's divinity was thought to be his golden thigh, a description that perhaps referred to a birthmark. Iamblichus tells of Abaris, a Hyperborean priest or druid, who was returning to his home in the north after a fundraising mission for his temple. The Hyperboreans were the ancestors of Celtic tribes and worshippers of Apollo. On his way through Italy, Abaris saw Pythagoras and became convinced by his appearance that he was none other than Apollo himself. He offered Pythagoras the most precious thing in his possession, a sacred arrow said to have belonged to Apollo, like the ones that killed Python. The arrow, Abaris claimed, had magical powers: whenever he had encountered obstacles on his travels, such as impassable rivers or mountains, the arrow had enabled him to fly across. He had used it also to stop epidemics and to purify Sparta of a mysterious toxin that was poisoning the city (perhaps toxic gases rising from Mount Taygetus).

Pythagoras accepted this magical arrow without any hint of surprise, "as if he was in reality a God himself."[10] He took Abaris aside, showed him his golden thigh to prove that Abaris was not mistaken, and explained that "he had come for the purpose of remedying and benefiting the condition of mankind, and that on this account he had assumed a human form, lest men being disturbed by the novelty of his transcendency, should avoid the discipline which he possessed."[11]

No written works by Pythagoras have survived. We know that he was born on the island of Samos, in the Aegean Sea, sometime in the sixth century B.C. In his life, he travelled and studied extensively: with the mathematician Thales of Ionia (who forecast the yields of harvests, and predicted an eclipse of the sun in 585 B.C.), the Phoenician sages of Syria, and the high priests of Egypt. He stayed in Egypt until the Persians invaded and he was taken to Babylon. He spent a further several years in the capital of Mesopotamia before finally returning to Samos.

In Samos he set up a school, known as the semicircle, to study philosophy and hold political meetings. He lived outside the city in a secluded cave, where he carried out his mathematical research. As his popularity and reputation grew, the citizens of Samos began to draw on his help with city affairs, intruding on the privacy and calm that he required for his studies. At about the age of forty, Pythagoras left Samos and went to Croton, in southern Italy. There he formed a new, secretive society.

THE MOST PERFECT NUMBER

As both teacher and spiritual leader, Pythagoras attracted hundreds of students. Those in his inner circle, both men and women, were known as *mathematikoi.* To join the commune, they had to give up all personal possessions, follow a strict vegetarian diet and ascetic lifestyle, and study five years under a vow of silence. Pythagoras explained that the aim of these privations was to train the applicant's power of reason: "Excess brings lust, intoxication and uncontrolled emotions, which drive men and women into the abyss. Greed brings envy, theft and exploitation. These thickets, which choke the soul, must be cleared out by systematic discipline, as if with fire and sword. Only when reason is liberated from such evils are we able to implant what is useful and good within the soul."[12]

The *mathematikoi* were the hard-core Pythagoreans, the true priests of Apollo. They could quit the arduous program whenever they wanted, and recover all the material goods they had donated, times two. But if they did, a monument was constructed to them as for a burial, and they were regarded as dead; every time a Pythagorean passed them in the street, he would act as if they had never met.

The outer circle were known as *akousmatics*. They lived in their own houses, kept their possessions, were allowed to eat meat, and visited the society only during the day. However, they were not allowed to see Pythagoras, and were not taught the cult's inner secrets. When the *akousmatics* attended lectures, they sat in the back, separated from the master by a screen. They were never shown mathematical proofs, and instead had to accept the results *ipse dixit,* because Pythagoras said they were so.

Life in the commune adhered to a strict routine. Solitary or group walks were followed by lectures on astronomy, music, or mathematics; corrective counselling; and exercise sessions similar perhaps to Tai Chi or yoga. Some of the exercises might have been of Pythagoras's own devising; while in Samos, he had turned the athlete Eurymenes into an Olympic champion by making him follow an arduous training regimen. Lunch was bread and honey or honeycomb; dinner was vegetarian. In the evenings, Pythagoras would give lectures. These were typically attended by at least 600 people, with the *mathematikoi* at front and everyone else shielded by the screen.

One of the topics for the evening lectures was no doubt foretelling the future. Pythagoras had studied under Thales and was said to have surpassed his mentor in the art of prognostication. Like Thales, he is said to have been able to predict eclipses, harvests, and earthquakes, and perhaps through his influence with Apollo, could halt epidemics and calm storms. He taught many systems of prediction, such as the reading of entrails or listening to oracles. But for him, the highest form of prediction was divination through numbers, which Pythagoras thought connected more closely with the "celestial numbers of the gods" than other methods.[13] One of his students, Empedocles, became known as Alexanamos, or "Averter of Winds," for being able to predict and control the weather. (His modern counterpart is the U.S. evangelist Pat Robertson, who claims to have used the power of prayer to steer the course of hurricanes.) Just as Apollo's arrow had enabled Arabis to dart across landscapes without needing to traverse mountains or rivers, so the magic of numbers allowed the Pythagoreans to dart through time and foresee future events without having to wait for them to happen.

The details of how this system of numerical prediction worked remain unknown, since the group was obsessed with secrecy. According to Iamblichus, "Their writings and all the books which they published were not composed in a popular and vulgar diction, so as to be immediately understood, but in such a way as to conceal, after an arcane mode, divine mysteries from the uninitiated."[14] Rather than rely on written records, the Pythagoreans were trained to improve their powers of memory; each morning before arising, for example, they would recount to themselves the exact events of the previous day. We would know little of Pythagoras's teachings if it weren't for the writings of subsequent philosophers, such as Plato and Aristotle. This secrecy certainly also added to Pythagoras's mystique.

For the Pythagoreans, numbers were much more than a tool for prognostication. Rather, they were what united the reason of man with the workings of nature. Each number was a kind of mystical entity with its own special properties. By understanding these properties, man could gain insight into the workings of the world, see into the future, and become closer to the gods.

The monad represented the initial unified state from which the universe was created, and was associated with divine intelligence. The division of the monad into the dyad, the number two, symbolized polarization: unity became duality. The dyad therefore signified mutability, or the ability to change appearances, and also unlimited excess, conflict, and indeterminacy—all negative qualities in a commune where applicants were selected for their ability to control anger and passion. "Lamentations, weepings, supplications, entreaties were considered abject and effeminate and neither gain, desire, anger, ambition nor anything of a similar nature became the cause of dissension among them."[15] The number three, the triad, enabled all things with a beginning, a middle, and an end, or a past, a present, and a future. It was the number associated with prophecy, as in the tripod at Delphi. Number four, the tetrad, represented completion, as in the four seasons that make up a year. The greatest and most perfect of all numbers was the decad, ten. Just as the first four numbers sum to ten, the decad was also the sum of the laws of nature. The following arrowhead arrangement of ten dots, known as the *tetractys,* was used by the Pythagoreans as a sacred symbol:

RIGHT VS LEFT

The dyad represented the division of the universe into two groups. Table 1.1, a list of ten pairs of antitheses, was compiled by the Pythagoreans in reference to the decad and documented in Aristotle's *Metaphysics*. These antitheses were believed to represent fundamental organizing principles of the universe.

TABLE 1.1

Limited	Unlimited
Odd	Even
One	Plurality
Right	Left
Male	Female
At Rest	In Motion
Straight	Crooked
Light	Darkness
Square	Oblong
Good	Evil

Pythagoras believed in reincarnation and claimed to be able to remember his past lives. He once rescued a dog from being beaten on the street and told the owner that he could tell by the animal's cries that it was the soul of his late friend Abides. Through repeated incarnations, the Pythagoreans believed that they could choose limited over unlimited, light over darkness—the first column over the second—and thus achieve divinity.[16]

Why the Pythagoreans chose these particular items for their list of opposites is unclear (though we explore some possible reasons later). It is interesting to compare it with the following lists (on page 30), which are from very different sources.

TABLE 1.2

Left Brain/Right Side	Right Brain/Left Side
Intellect	Intuition
Abstract	Concrete
Analytic	Holistic
Rational	Intuitive
Objective	Subjective

TABLE 1.3

Yang	Yin
Odd	Even
Conscious	Unconscious
Right Side	Left Side
Masculine	Feminine
Aggressive	Yielding
Light	Darkness
Reason	Emotion

TABLE 1.4

Physical Science	The Humanities
Hard	Soft
Determinism	Free Will
Reason	Feeling, Emotion
Objective	Subjective
Quantity	Quality
Specialism	Holism
Prose	Poetry
Male	Female
Clarity	Mystery

The list in table 1.2 was the result of so-called split brain studies, conducted with patients who had been disabled by extremely severe epileptic seizures.[17] The human brain is divided into two hemispheres, with the left controlling the right side of the body and vice versa. In a last attempt at therapy, connections between the two hemispheres were severed to stop the seizures spreading across the brain. While the treatment succeeded in controlling the seizures, it effectively isolated the two sides. Through a series of experiments, the researchers attempted to determine the functions of each hemisphere. The left brain, they came to believe, is associated with abstract, rational thinking, and the right brain with holistic and intuitive modes of thought. In a healthy brain, the two sides work in concert, so it is never possible to cleanly separate their functions.

Table 1.3 is from the I Ching, or Book of Changes.[18] From these tables, one might deduce that Pythagoras was a left-brain (right-hand) kind of guy, more yang than yin. As Iamblichus wrote, "The right hand he called the principle of the odd number and is divine, but the left hand is the symbol of the even number and of that which is dissolved."[19] Apollo was the god of reason, and one of the commune's aims was to elevate rational, objective reasoning over subjective, emotional behaviour. (The preference for the right hand has continued in our language—the word "sinister" is from the Latin for left.) Table 1.4 is from a longer list compiled by the philosopher Mary Midgley, who wrote in 1985 that the instruction to keep with the items in the first column "has for the last century usually been issued to English-speaking scientists with their first test-tube and has often gone with them to the grave."[20]

Science has changed a great deal since the time of Pythagoras, but the emphasis on using reason and analysis to provide hard, fixed solutions for particular, specialized problems has remained the same. The development of quantum physics, which revealed the wavelike properties of matter, along with attempts to adopt a rounder, more holistic perspective, as in systems biology, has softened this distinction. But scientists on the whole are still squares, not oblongs, and the idea, so prevalent in science, that complex phenomena should be reduced to simple ones is Pythagorean. As we will see, this tendency to drive on the right has been both the strength and the weakness of scientific forecasting.

MUSIC OF THE SPHERES

Like Apollo, who was frequently portrayed with a lyre, Pythagoras was a musician. He believed that music had healing powers and could be used to calm the soul. A powerful proof of the importance of numbers was the discovery, attributed to him, of their role in music. A string on a lyre, when plucked, will give a particular note. Fretting the string at a position halfway down gives a note differing by an octave; a third of the length down gives a musical fifth; and one quarter the length a musical fourth. Use a different string, or an electric guitar, and the same relationship holds.

Pythagoras realized that the relationship between pleasing notes was all a question of numbers. And if music, one of the most expressive art forms, could be reduced to numbers, then so perhaps could everything else. In the Pythagorean cosmos (a word he invented), the stars, planets, moon, and sun were contained in nested, concentric, transparent spheres, all of which rotated around the earth according to a cosmic harmony, which Pythagoras called the Music of the Spheres. He argued correctly that the earth itself was a sphere that caused night and day by its revolution, and that the seasons were the result of the angle of the earth's axis with the sun. Even time itself was cyclical, repeating itself once every Great Year. This was the period it took the sun, moon, and planets to return to the same configuration, estimated by ancient astronomers to be around 10,800 years.[21]

Pythagoras is credited with a number of mathematical discoveries, including the properties of what are now called the Platonic solids—the pyramid-shaped tetrahedron, cube, octahedron, dodecahedron, and icosahedron. (Every face of these polyhedron figures is identical, and remarkably only five exist.) However, he is best known for his famous theorem, which states that in a right triangle, the square of the side opposite the right angle equals the sum of squares of the other two sides. While the Egyptians and Babylonians were probably aware of this relationship well before Pythagoras, at least for certain triplets, the Pythagoreans appear to have been the first to generalize the concept. Just as the theory of musical harmony applies to any instrument, the Pythagorean theorem applies to any right triangle. As the Pythagoreans understood, the power of mathematics comes from knowing that a single law holds in all cases—from reducing the plurality to one. The theorem is still one of the most important results in mathematics, and it's used in everything from engineering to nuclear physics.

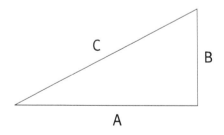

FIGURE 1.2. The theorem of Pythagoras: $A^2 + B^2 = C^2$.

The Pythagoreans believed, almost as an article of religious faith, that the world was made up of positive integers and their ratios, such as the fraction ¾, which are called rational numbers. Ironically, Pythagoras's theorem about triangles led to the discovery of numbers that cannot be expressed as a ratio. A right triangle with two sides both equal to one unit has a hypotenuse (from the theorem) of the square root of two. Hippasus, one of the Pythagoreans, showed that the root could not be expressed as a ratio of two integers. In other words, it was irrational. His comrades could not accept that such a number existed: it was as if someone had found a bum note in the music of the spheres. Hippasus made the mistake of publicizing the results openly, "to the profane and to those . . . without disciplines and theories."[22] He died shortly afterwards under mysterious circumstances. It was said that "the Divine Powers were so indignant that he perished in the sea."[23] (See notes for a proof that Hippasus was right.[24])

It is strange that numbers that cannot be expressed as a ratio of whole numbers are called irrational, as if they were in some way deviant; there are far more of them than there are rational numbers, just as there are many more pitches of sound than those found on a keyboard. In fact, if you could choose only one number by throwing an imaginary dart at the interval from zero to one, the chances of hitting a rational number are zero.[25] You might aim to hit ½, but you'd actually get some irrational number with an endless sequence of digits, like 0.5083428 . . .

The Pythagorean commune grew in both size and power, to the point where it exerted considerable influence over Croton and the surrounding area. It is even believed that Pythagoras became the local "master of the mint," bringing the first metal coinage to the region.[26] Eventually, though, this rational society became the victim of seemingly irrational forces. People—especially those who had been excluded from membership—began to speak against the secretive and elitist group. The citizens started to harass the Pythagoreans in the streets. When a number of them assembled at the house of Milon, an Olympic athlete, a mob surrounded them and set the house on fire; only two escaped the conflagration. In another incident, forty members of the group were attacked and killed. Pythagoras himself managed to escape, and probably died in exile. Even Apollo's arrow was no protection against the madness of crowds; but the demise of the Pythagoreans marked only the first stage in the development of numerical prediction. As the novelist and philosopher Arthur Koestler wrote of Pythagoras: "His influence on the ideas, and thereby on the destiny, of the human race was probably greater than that of any single man before or after him."[27] He didn't just predict the future; he also helped define it.

THE ACADEMICS

Since the Pythagoreans didn't believe in recording their methods, our accounts come mostly from future documenters. One of these was a man born, it is said, with the name Aristocles. His mother and father came from famous, wealthy families in Athens. His uncle was a friend of the philosopher Socrates. Perhaps because of his physical bulk—he was a trained wrestler—or the width of his forehead, he was usually known by his nickname, which roughly translates to "the broad." He was Plato.

Like Pythagoras, Plato was a man of many talents. He knew the arts of politics, philosophy, and war. He was also a playwright. According to Diogenes, he began his career as a writer of tragedies. After hearing Socrates talk, however, he gave up on the theatre, and even set fire to a play that he had been planning to enter into a drama competition. Instead, he began pouring his creative energy into the writing of philosophical dialogues. While these weren't plays, they did show his enormous skill at crafting entertaining dialogue. They also often blur the line between fiction and non-fiction. It is hard to know whether the dialogues represent things that were actually said or are a fictionalized account. Or whether they even represent what Plato himself thought on a subject.

An example of this ambiguity is to be found in Plato's *Defence of Socrates*. Socrates, the son of a sculptor, was another servant of Apollo who dedicated his life to understanding the causes that underlie the universe. This quest took on a new form after his friend Chaerephon visited the Delphic Oracle and asked whether there was anyone who was wiser than Socrates. The Pythia replied that there was not. Since Socrates was unaware of any wisdom within himself—he often joked that "the only thing I know is that I know nothing"—he interpreted this to be a mission from Apollo to visit those who claim to be wise and discover their wisdom. He therefore went to poets, artisans, and statesmen across the land, but after closely questioning them, he realized that they were not wise at all. "And so I go my way, obedient to the god, and make inquisition into the wisdom of anyone, whether citizen or stranger, who appears to be wise; and if he is not wise, then in vindication of the oracle I show him that he is not wise; and this occupation quite absorbs me, and I have no time to give either to any public matter of interest or to any concern of my own, but I am in utter poverty by reason of my devotion to the god."[28]

Needless to say, this attitude annoyed a lot of people. In 423 B.C., Aristophanes wrote a comedy called *The Clouds,* in which the main character, also called Socrates, worships clouds and other natural phenomena rather than the gods. The play was produced in a competition at the Great Dionysia. It came third out of three plays but was published a few years later. The Greeks were avid theatre-goers, and the play turned Socrates into first a figure of fun, then a tragic anti-hero.

Twenty-four years after *The Clouds* was produced, Socrates found himself on trial, accused of believing in new divinities and corrupting the youth. Perhaps his accusers were confusing the character in the play with the real person. Plato's *Defence of Socrates* is a dramatic account of how the philosopher defended himself against the citizens of Athens. It is not known whether it is a verbatim transcript of what Socrates said, a heavily edited version, or a fictionalization that is really Plato's defence of his friend and mentor. Again, the story and the reality are hard to separate.

In any case, according to Plato, the jury was not convinced by Socrates' case. The prosecutors were seeking the death penalty, but Socrates was given the chance to offer an alternative. He first suggested free meals for himself for life. This didn't go down well, so he suggested a nominal fine of one mina. As Socrates dug a deeper and deeper hole for himself, his friends, including Plato, offered to pay a more substantial fine. But it was to no avail, and Socrates was put to death by poisoning with hemlock.

The death of Socrates was not in vain, for it had a huge impact on Plato. Sick of the politics of Athens, he travelled to Egypt, Sicily, and Italy. It was in Italy that he learned of the work of Pythagoras and met his disciples. From them, according to the scholar G.C. Field, he formed the idea "that the reality which scientific thought is seeking must be expressible in mathematical terms, mathematics being the most precise and definite kind of thinking of which we are capable."[29] The only way to overcome the ignorance that Socrates had exposed was with numbers.

When Plato returned to Athens, around 387 B.C., he established what became the longest-running learning institution in the history of mankind—the precursor to today's universities. The Academy, so named because the land belonged to a man called Academos, was dedicated to research and instruction in philosophy and science. Over the door was written, "Let no one unversed in geometry enter here." Plato's concentration on precise definitions, clear statement of hypotheses, and rigorous proofs of mathematical conjectures all prepared the ground for the major mathematical developments of ancient Greece, which underpin modern science. The Academy survived more than 900 years, until the Christian emperor Justinian, claiming it was a pagan establishment, closed it down in 529 A.D.

Students at the Academy would spend ten years studying the sciences of astronomy and mathematics, then five years studying dialectic (the art of posing and answering questions). Plato believed that dialectic was the path to wisdom, and through his dialogues, he contributed to the theory of arts from poetry to epistemology. He taught that material objects were imperfect versions of underlying forms, which existed in a static way, independent of time and space. An example of a perfect form was a mathematical object such as a line. A material manifestation of a line, such as a line drawn in the sand, was only a flawed reproduction of the real thing, like a poor photocopy. Every object had an associated form, which it yearned to be but could never reach. The plurality of different tables, for example, all aspired to the one true table. To Plato, the ultimate reality was not the chaotic, imperfect world that we see and hear and taste, but rather the abstract, eternal world of pure forms. Our world was just a blurred shadow of the real thing.

MATHEMATICAL BIOLOGY

In 430 B.C., the citizens of Athens were struggling with the real-world problem of infectious disease. Thucydides gave a graphic description of the plague that was afflicting the city. People were first attacked by "violent heats in the head, and redness and inflammation in the eyes,"[30] along with sneezing, hoarseness, and a cough.

"Discharges of bile of every kind named by physicians ensued, accompanied by very great distress." The skin was "reddish, livid, and breaking out into small pustules and ulcers. But internally it burned so that the patient could not bear to have on him clothing or linen even of the very lightest description; or indeed to be otherwise than stark naked. What they would have liked best would have been to throw themselves into cold water; as indeed was done by some of the neglected sick, who plunged into the rain-tanks in their agonies of unquenchable thirst; though it made no difference whether they drank little or much." In most cases, the disease proved fatal after seven or eight days. Some, like Thucydides himself, survived but were often maimed or blinded.

Near the height of the plague, a delegation was sent to Delphi to ask the oracle how it could be stopped. The oracle's reply was that the altar of Apollo on the island of Delos, which was in the form of a perfect cube, should be doubled in size. In response, the delegates arranged for each edge of the cube to be extended by a factor of two; however, this increased the volume not by two but by eight. The oracle announced that Apollo—whose arrows were believed to cause plague sores—had been angered by this sloppy arithmetic, and indeed the outbreak grew worse.[31] Plato was consulted. He told them, "The god has given this oracle, not because he wanted an altar of double the size, but because he wished in setting this task before them, to reproach the Greeks for their neglect of mathematics and their contempt of geometry." Only the magic of number could defeat the plague.

Perhaps he was right, because soon the plague began to ease, though not before claiming about a third of the population of Athens. The problem of how to double the cube didn't go away, though. The Athenian mathematicians, at least those still surviving, believed that all mathematical problems could and should be solved using only a compass and a ruler, tools that corresponded to the perfect forms of circles and straight lines. Anything that couldn't be expressed in these terms was out of bounds. Just as the Pythagoreans eventually encountered a problem that could not be solved by rational numbers, however, the Athenians found that many of their problems could not be solved with these two tools alone. They could not construct a square with the same area of a given circle or double a cube or trisect an angle into three equal angles. Their insistence on static forms stood solidly in the way of progress.

Eventually, solutions for all these challenges were arrived at using so-called mechanical curves. These needed to be traced out by sliding lines around a point. Because they introduced the idea of change and motion—a bad thing in Pythagoras's list—they were not considered to be real geometry.

THE GREEK CIRCLE MODEL

To Plato and other philosophers of his time, the universe was not a place of chaotic flux and change but a kind of endlessly repeating cycle. Like Pythagoras, Plato thought that time moves in circles. The future had already been determined, and it was the past. Because events did not occur randomly, but were known in advance, it followed that the future could be predicted. The logical place to start was up above, in the heavens.

The exemplars of circular, repetitive motion were the stars and the planets, which were believed to move in perfect circles around the earth. Indeed, careful observations of the stars showed that they did move in a circular fashion. The planets—in particular Mars, Jupiter, and Saturn—were more tricky. Their path around the night sky involved a fair amount of wandering (the word "planet" is from the Greek for wanderer) and even backtracking. It seemed they had a life of their own. For example, Mars advanced around the sky from west to east, completing a revolution in about 780 days; but partway around it would stop, backtrack a little, and then resume its forward motion. And 780 days later, it would do the same thing.

The wanderings of the planets seemed incompatible with simple circular motion, but Plato's associate Eudoxus managed to come up with a model for the universe that captured such effects.[32] Imagine the earth surrounded by a huge crystalline sphere that contains the stars. The sphere rotates around us once per day. Inside this sphere is a separate transparent sphere that contains the sun. It too rotates around the earth once a day, but it also rotates annually at an approximate 23.5-degree angle to the line joining the centre of the earth to the North Pole. The angle accounts for the seasons, since half the year one side of the globe will receive more sun, while the other half of the year it receives less (as shown in figure 1.3 on page 42). This much was simple; the movements of the planets were more complicated. These were also modelled by spheres, whose axes of rotation were fixed to other spheres (which could themselves rotate). The resulting nest of twenty-seven rolling spheres was capable of producing highly complex motion. With carefully selected rates and angles of rotation, the model could adequately represent the motion of the heavens.

This model, which I will refer to as the first Greek Circle Model, was an amazingly ingenious geometrical accomplishment, and it can be viewed as a direct precursor to the mathematical models that are used today to simulate physical systems. Of course, it was purely descriptive, and was based on a hypothesis of circular motion, as opposed to rigorously derived physical laws of motion. The fact that it worked quite well as a model of the universe is a poignant reminder that a model that can be made to fit the data isn't necessarily an accurate representation of reality.

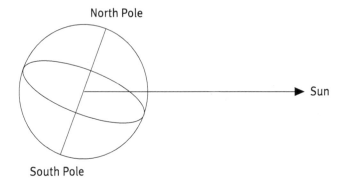

North Pole

Sun

South Pole

FIGURE 1.3. The angle of the earth relative to the sun means that each hemisphere gets more sun in half of the year (its summer) and less in the other (winter). The angle actually varies from 21.8 to 24.4 degrees, returning to centre about every 42,000 years, owing to slight wobbles in the planetary system.

THE WORLD'S TUTOR

The Academy was the elite institution, the Ivy League or Oxbridge of its time, and many of Plato's students went on to make major contributions to Greek science and philosophy. His star student was Aristotle, who stayed at the Academy for twenty years, from the time he was eighteen until Plato's death in 384 B.C. Aristotle then took a job with King Philip of Macedonia, tutoring his son Alexander for three years. We are familiar with the work of Aristotle and Plato largely because of Alexander the Great. When Alexander went to Delphi to obtain his oracle, the Pythia refused. He kept insisting, and even threatened her with force. Finally, she told him, "You can do what you like."[33] This was like telling George W. Bush not to hold back so much, and Alexander went on to conquer the Middle East, Persia, and Egypt, as well as parts of Afghanistan, Central Asia, and India. The library in one of the cities named after him—Alexandria, in Egypt—became the major repository of

Greek knowledge, and eventually cemented Aristotle's position as tutor to much of the world.

In 355 B.C., Aristotle returned to Athens and set up his own institute, called the Lyceum after the old temple of Apollo in which it was located. Like the Academy, the Lyceum taught a range of subjects, such as politics, ethics, and science. While Plato was fascinated by the abstract properties of forms, Aristotle's science was more grounded in observation of physical and natural phenomena. For example, he collated descriptions of about 500 different types of animals, many of which he dissected. In Raphael's painting *The School of Athens,* Plato is shown gesturing to the heavens while Aristotle is lowering his hand to the ground, as if to bring the theorist back to earth.

Aristotle viewed material substance as bestowed with a kind of life force with its own wants and desires. He believed that all substances were composed of the four elements—earth, water, air, and fire. The tendency of earth is to sink strongly down. Water trickles down less strongly, while air rises and fire positively springs to the sky. An air bubble in water will rise upwards because air "wants" to be higher than water. Motion, therefore, occurs either because an object wants to find its own level or because it is pushed. A full explanation of any object had to take into account its final cause, the purpose for which the thing existed. The stars in heaven were made of the fifth element, called ether, the lightest of all, and moved in a circle, which was the figure of eternal motion. Earth had to be in the middle of the universe, because it was the heaviest thing around.

In this teleological view of the world, the earth itself was a kind of organism. Natural phenomena such as earthquakes, winds, or even meteors were the result of the planet's "windy exhalations." The son of a doctor, Aristotle constantly drew comparisons between the earth and human bodies. He believed that tremors or spasms were caused by a kind of wind within the body, and that earthquakes were caused by a similar wind, but on a larger scale.

Perhaps Aristotle's most significant contribution to science was his axiomatic development of logic. In his work *Prior Analytics,* he proposed his syllogistic form of argument—the ultimate in linear, left-brain thinking—which consisted of two premises and a conclusion. His gloomy but hard-to-counter example was:

(i) Every Greek is a person.
(ii) Every person is mortal.
(iii) Every Greek is mortal.

(One imagines that the first spinoff from the Lyceum was a life-insurance company.) This systematic, logic-based approach to science laid the foundation for Euclid's development of geometry and helped establish what became known as the scientific method.

While Aristotle's work in biology has been much admired, his theories in physics were less reliable. He postulated two laws of motion. The first was that the heavier an object is, the faster it will fall. The second was that the speed of fall decreases with the density of the medium—so, for example, a stone will fall more slowly in water than it will in air. Curiously, while Aristotle made detailed observations of many biological specimens and natural phenomena, he didn't verify his theories of physics. It was left to Galileo, nineteen centuries later, to actually drop stones off buildings and disprove Aristotle's first law.

GREEK CIRCLE MODEL, VERSION II

At the Lyceum and elsewhere, astronomers continued to improve the Greek Circle Model. The original version of Eudoxus captured both the daily and yearly cycles of the sun and the general motion of the planets, but it didn't match some of the details. In particular, it was known that the seasonal motion of the sun, as measured by the time between solstices, was not uniform. This was repaired by adding more spheres of motion. The final model, which was presented by Aristotle and accounted for the motions of all the visible planets and the moon, included no fewer than fifty-five concentric spheres. It matched the observed movement of the planets around the sky and consisted solely of circular motion, which was the only type that could occur in the ether.

After Aristotle's death, a mathematician called Aristarchus of Samos proposed the novel theory that the earth revolved around the sun, rather than the other way round. The stars do not rotate around the earth, he suggested, but stay in their positions an enormously far distance away. (The large distance was required so that the stars appear not to move relative to one another as the earth rotates.) Perhaps because it was incompatible with the views of Aristotle, the idea did not catch on.

The serious mathematicians continued to tinker with the Greek Circle Model. Around 150 A.D., Ptolemy of Alexandria put the finishing touches on a new version. It included some tweaks of Aristotle's model, and some major changes. It was known that the size of the moon and the brightness of the planets tended to vary, which suggested that their distance from the earth changed with time. The most straightforward way to address this would have been to adopt non-circular motion, but again this would have contravened dogma. Ptolemy wrote: "Our problem is to demonstrate, in the case of the five planets as in the case of the sun and moon, all their apparent irregularities are produced by means of regular and circular motions (for these are strangers to disparities and disorders)."[34] He achieved this by incorporating a new type of circular motion, first proposed by Hipparchus, known as epicycles—that is, circles within circles. In Ptolemy's model, the planet rotates around a small circle, which in turn rotates around the earth (as shown in figure 1.4). Its distance from the earth therefore varies, as does the rate at which the planet moves. Working all this out required the invention of trigonometry, which some attribute to Hipparchus.

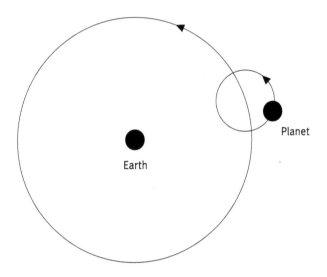

FIGURE 1.4. In the Greek Circle Model, Version II, planets move in epicycles—circles within circles.

With its cycles, epicycles, and even eccentric epicycles (whose centres were slightly offset from the main cycles), the entire model was even more complicated than Aristotle's. By insisting at a basic level on the Pythagorean simplicity of circular motion, the model effectively exported the system's complexity to a higher level. The extreme flexibility in the model meant that it could be made to closely match the observational data. Ptolemy wrote up his results, which included scores of tables detailing the motions of the heavens, in the *Almagest* (from the Arab *al-majisti,* meaning "the greatest"). Its publication marked the transformation of astronomy from a theoretical pursuit into a predictive science. Just as Apollo's arrow had allowed the druid Abaris to fly across obstacles, the mathematically based Greek Circle Model allowed the Greeks to forecast the future by looking it up in a book.

STARGAZERS

It may seem to the modern reader that predicting the motion of the stars and planets, while useful for chronology and navigation, has little to do with practical forecasting in weather, medicine, or economics. Indeed, for most people living in urban areas, the heavens are rendered all but invisible by light pollution. However, prior to the invention of the electric light bulb (or for that matter, competition from the stars on TV), the night sky played a rather more important role in people's lives. The Greeks, like others before and after them, believed that life here on earth was in harmony with the heavens. The arrival of a comet could herald drought, famine, or political upheaval. The positions of the moon and the planets could be interpreted to make predictions about the weather and the harvests (as described in Ptolemy's astrological work, *Tetrabiblos*). Astrology was also used to determine the optimal timing for medical interventions, and to cast horoscopes of newborn children.

Astrology is rooted in the belief that the soul comes from and is united with the heavens. In Plato's dialogue *Phadros,* Socrates (whose violent death was predicted to him by a Syrian astrologer) says that "soul, considered collectively, has the care of all that is soulless, and it traverses the whole heaven, appearing sometimes in one form and sometimes in another. . . . The whole, compounded of body and soul, is called a living being, and is further designated as mortal."[35]

The first astrological text, from the second millennium B.C. (the Old Babylonian period), offers the following advice, inferred from the state of the moon and the sky at the start of the New Year:[36]

1. If the sky is dark, the year will be bad.
2. If the face of the sky is bright when the New Moon appears and [it is greeted] with joy, the year will be good.
3. If the North Wind blows across the face of the sky before the New Moon, the corn will grow abundantly.
4. If on the day of the crescent the Moon-God does not disappear quickly enough from the sky, quaking [presumably some disease] will come upon the Land.

More detailed and sophisticated astrological forecasts based on the zodiac rely on knowledge of the positions of the sun, the moon, and the planets at particular times of day, like the exact time of a child's birth. Since direct observations are impossible during the day or when it is overcast, astrologists needed to first predict the state of the heavens. As a result, astronomy and astrology were viewed throughout much of history almost as separate branches of the same science.

Astrological forecasters still play a role, of course—even in the highest circles of business and politics. Nancy Reagan consulted with an astrologist to help avoid assassination attempts against her husband after the one on March 30, 1981.[37] In a dangerous, chaotic world, any predictions can sometimes seem better than none at all. And unlike many other methods of prognostication—say, the reading of animal entrails (hard to imagine that in the White House)—astrology is based on a defined, apparently rational method, which to many gives it an aura of credibility.

METHODS OF DIVINATION

Traditional techniques of foretelling the future include:
Stars and planets (astrology)
Rolling dice/drawing lots (cleromancy)
Tarot cards (cartomancy)
Palm reading (chiromancy)
Crystal balls (crystallomancy)
Shape of head (phrenology)
Atmospheric conditions (aeromancy)
Dreams (oneiromancy)
Animal entrails (haruspicy)
Moles on the body (moleosophy)
Lightning and thunder (ceraunoscopy)
Smoke and fire (pyromancy)
Flight of birds (ornithomancy)
Neighing of horses (hippomancy)
Tea leaves or coffee grounds (tasseomancy)
Passages of sacred texts (bibliomancy)
Numbers (numerology)
I Ching
Guessing
To which we can add:
Mathematical models (meteorology/biology/economics)

Whether distant planets control the hand of an assassin is certainly debatable, but subtle astronomical rhythms have been implicated in the timing of the ice ages (as will be discussed later). Perhaps the biggest reason for wanting to predict the movements of the stars and planets, though, is to prove that unlike most other things here on the ground, they do move in a predictable fashion. As Karl Marx said, "Mankind always sets itself only such problems as it can solve."[38] The Greek Circle Model was a first step in the tradition of using mathematics to make predictions about the physical universe.

During his own time, Aristotle was not regarded as highly as Plato. His work, much of it compiled from lecture notes, was probably not even published until hundreds of years after his death, and it gained prominence in Christian Europe only when translated into Latin in the twelfth and thirteenth centuries. From that point on, however, he became the almost unquestioned authority, his word the gospel truth.

Aristotle, therefore, both greatly advanced and inadvertently retarded the progress of science. On the one hand, he helped develop the system of rational, logical thinking—the legacy of Pythagoras—that culminated in modern scientific models. On the other hand, his enormous, if belated, prestige seems to have stilted the application of those methods to the natural world. In the Middle Ages, his texts had become as solid as a star in the firmament, and the Greek Circle Model had taken on the status of a perfect form. It was as if he opened his own eyes, only to close those of other people for hundreds of years.

In many ways this is ironic, since Aristotle was no blind follower of authority himself. He criticized earlier philosophers for defending their fixed ideas, their mental models of the universe, against all the facts.[39] It took Tycho Brahe and other scientists of the sixteenth century to show that even the stars in heaven aren't as immutable as they appear.

which was born in Pythagoras's cult and developed by the ancient Greeks, entered in the Dark Ages into what might be called a static phase, where little progress was made for hundreds of years. Many advances were made in mathematics—one of the biggest being the introduction of Arabic numerals, which considerably simplified computation—but in the fifteenth century, the state-of-the-art model of the cosmos was still the Greek Circle Model. Part of the problem was the "lock-in" phenomenon, which gives ideas or technologies that first gain a foothold a kind of historical inevitability, as if they were God-given truths (rather like the Windows operating system on my computer). When new ideas do eventually emerge, they tend to spring out of a number of places at around the same time, as if they can be repressed no longer.

The Greek Circle Model had originally been based on the belief that earth is the heaviest and basest of the four elements, and therefore has a tendency to sink down. Since the planet and the cosmos itself were spherical, down meant towards the middle. The stars and planets, which are made up of ether, the lightest and most pure element, would naturally tend to spin around this dense, immobile core. Our earth was not so much at the centre of the universe as it was at the bottom. Perfection was up there in the heavens, not down here on the ground.

When the Greek Circle Model was adopted by men like St. Thomas Aquinas to build a Christian cosmology, it came to represent not just a physical model of the cosmos but a theological one. According to the Christian view, humans were made in the image of God. It followed that our place in the universe was at the centre, with everything revolving around us. The Greeks' original ordering of the universe was therefore inverted: the perfect stars of God's creation were here on earth. The Greek Circle Model became the bible for the heavens—questioning it was like questioning the whole foundation of the Christian model of reality.

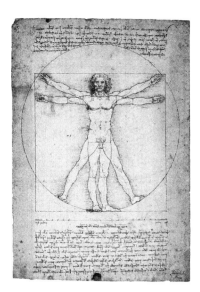

FIGURE 2.1. Leonardo da Vinci, *The Proportions of the Human Figure (after Vitruvius), c.* 1490.

The momentum for change—which would lead from the Dark Ages to a huge blossoming of science in the Enlightenment of the seventeenth and eighteenth centuries—came not from astronomers but from Renaissance figures such as Leonardo da Vinci. Leonardo—who was lauded by the poet Baldassare Taccone as a "geometer" as well as an artist, and spent as much time working on military inventions and engineering projects as he did on canvas—described his desire for knowledge as follows: "Unable to resist my eager desire and wanting to see the great [wealth] of the various and strange shapes made by formative nature, and having wandered some distance among gloomy rocks, I came to the entrance of a great cavern, in front of which I stood some time, astonished and unaware of such a thing. . . . [T]wo contrary emotions arose in me, fear and desire—fear of the threatening dark cavern, desire to see whether there were any marvellous things within it."[2]

Renaissance artists like Leonardo created their works using both careful observations and mathematical theory (such as perspective techniques), and therefore elevated art to the same exalted level occupied by the classical, physical sciences. His 1490 work on human proportion (shown in figure 2.1 on page 53) is based on the only ancient work on the subject to have survived. The original, by Vitruvius, an architect and engineer during the time of the Roman Empire, positioned a man's hands and feet at the corners of a square inscribed inside a circle, with the navel at the centre. The result was a man whose appendages were unnaturally distended, as though striving to comply with the demands of classical geometry. Leonardo's version, which was informed by careful measurements that he carried out on actual people, adjusts the position of the square, as if to say that theory is great, but it has to agree with reality.

New World Order

The New World independently developed agriculture, civilization, astronomy, and methods of prediction. The Mayan predictive system was based on the ancient Long Count calendar. It prophesied, among other things, that December 21, 2012, would mark the beginning of a new cycle of time; some associate this with a great purification, when the earth will cleanse itself and many living things will die. So that's a good day to stay in.

The Aztec rulers of Yucatán, in pre-Columbian Mexico, predicted that 1519 would mark the birth of a blond, bearded deity known as Quetzalcóatl, or Feathered Serpent. That year, the Spaniard Hernán Cortés arrived with his army at what is now Veracruz. Cortés had blond hair and a beard. Believing their forecast had come true, the Aztecs greeted him as a god. It could have been the start of a great relationship, but unknown to Cortés, one of his men carried a different kind of serpent—smallpox. The Aztecs had no immunity; a quarter of their population is estimated to have died in the ensuing epidemic.

The impact of smallpox on the Incas was even stronger. The disease swept through their entire society within months, killing the emperor and much of the rest of the population. In 1636 it reached Lake Ontario, and by the end of that century millions of Native Americans had perished. The true conquistadors, the conquerors of America, were not men but microbes. (If smallpox had been a disease of the New World, instead of the Old, those age descriptions might have been different.) As discussed in Chapter 7, disease epidemics still sometimes come from where they are least expected.

The questioning, skeptical attitude soon worked its way into astronomy. In early 1543, the sixty-year-old Polish astronomer Nicolaus Kopernik, who preferred to be known by his Latinized name of Copernicus, lay on his bed in the castle turret where he lived, staring into a different kind of threatening, dark cavern: impending death. In his arms was an object of intense fear and desire, his freshly published work, *On the Revolution of Heavenly Spheres*. It contained a theory that he had been working on for thirty years, since he had received his doctorate in Italy. Like Aristarchus over a millennium before, Copernicus believed that the earth rotated around the sun, rather than vice versa. The universe might be based on circles, but we were not at their centre.

No doubt anticipating the reception his ideas would receive, he had delayed publication until near the end of his life. The manuscript he had sent to the publishers had the motto "For mathematicians only" and included the following pre-emptive strike against his critics: "Perhaps there will be babblers who, although completely ignorant of mathematics, nevertheless take it upon themselves to pass judgement on mathematical questions and, badly distorting some passages of Scripture to their purpose, will dare find fault with my undertaking and censure it. I disregard them even to the extent as despising their criticism as unfounded." He must have been shocked to find that in the version of the book in his arms, this defiant statement had been removed. The publishing process had been supervised by the Lutheran theologian Andreas Osiander, and he'd replaced the original preface with an unsigned letter claiming that the results of the book were intended not as truth but only as a more efficient method to calculate the positions of the sun and the planets.

The switcheroo by Osiander probably outraged the ailing Copernicus, but it did make his ideas more palatable to other astronomers. After all, his model avoided many of the tortuous complexities of the Greek Circle Model, and was therefore a useful device for computing orbits. However, because it still incorporated circular motion, it could not avoid the use of epicycles, and in fact employed even more than the Ptolemaic system (forty-eight versus forty).[3] And theology aside, few were willing to believe for real that the earth was spinning at high speed around the sun. At a time when rapid transit meant a fast and bumpy carriage ride, one would expect to be conscious of such motion.

THE OBSERVER

Almost thirty years after Copernicus's death, on the evening of November 11, 1572, a young Danish astronomer named Tycho Brahe emerged from his alchemical laboratory. Looking up into the night sky, he immediately noticed that something had changed. For the best part of ten years, he had studied the constellations; they were as familiar to him as a lover's face. Yet that night, almost straight above him in the constellation Cassiopeia, there shone a new star, as bright as Venus. Perhaps wondering if this was a side effect from inhaling the fumes of one of his alchemical experiments, he called his assistant to confirm. But there was no doubt: a star was born.

The sudden arrival of this star, or supernova (as they are now known), which remained visible for seventeen months, caused a panic across the continent. Some believed that it signalled the coming of a plague, and they were apparently proved right when there was an outbreak in Europe shortly afterwards. It also came as a shock to astronomers. The authorities on matters of the stars were the ancient Greeks, and if there was one thing the Greeks knew, it was that there was nothing new under the sun. Tycho, who at that moment decided to dedicate himself to astronomy instead of alchemy, carried out detailed observations of the new star. When published in 1574, they conclusively proved that the star was at such a distance that it was really an addition to the firmament—which, it seemed, was not so firm. He dealt a further blow to orthodoxy in 1577, when his calculations showed that a comet he observed on November 13 had passed through the realm of the planets, and therefore should have smashed into Aristotle's crystalline spheres, if they really existed.

Tycho, who had Latinized his name from Tyge, was born into nobility. His uncle Joergen, who was also his foster father, had saved King Frederik II of Denmark from drowning, then died himself of pneumonia. The king returned the debt by granting Tycho, whose brilliance had impressed him, his own observatory, called Uraniborg. It was located on the island of Hven (now Ven) in Copenhagen Sound, across the water from Hamlet's Elsinore. (Tycho was also a relative of Friedrich Rosenkrantz, who was immortalized in Shakespeare's play of 1600.) The island contained around forty farms and a village, and the residents became Tycho's subjects. Tycho himself designed the laboratory, based on buildings he had seen in Venice.

The observatory was stocked with the best instruments available, short of telescopes (which had yet to be invented).[4] The basement held an alchemical laboratory, in case Tycho needed a break from the stars. There were servants' quarters, a cell for any tenants who caused trouble, and a printing press, which was used for publishing both scientific works and his poetry. It was all set in a garden arranged on the classical forms of circles and squares. He soon established a reputation as a mediocre poet but the best naked-eye astronomer in Europe; he was able to plot the positions of celestial objects to better than two minutes (sixtieths of a degree). The stars had never before been pursued with such rigour, precision, equipment, and ambition. As one of his elegies read:

> Like blind moles, lethargic mobs see
> No more than earthly, perishable things.
> So very few Apollo grants to see
> The riches which Olympus hides away . . .
> More beautiful by far the goal they seek,
> For it is not a goal unknown to gods:
> Through mental force control the heaven's stars,
> Subject the ether to his conquering spirit.[5]

FIGURE 2.2. The observatory of Tycho Brahe at Uraniborg.[6] The main observatory building is at the centre.

Tycho knew of Copernicus's theory and was sufficiently curious to check it with observations. If the sun-centred model was true, then the position of the stars should change as the earth rotates around the sun. However, even with his best tools, Tycho could detect no such parallax shift. Either the earth was fixed or the universe was so incredibly large that the earth's motion around the sun was minor in scale. Finding this too hard to accept, he came up with a hybrid model that had the planets rotating around the sun, but the sun and the moon rotating around the earth, so the earth was still the centre of the universe.

Actually, Tycho came to think that *he* was at the centre of the universe. Scientists are often perceived publicly as being a humble group of truth-seekers, but the reality is sometimes different. As Iamblichus wrote, humility was not a virtue for the Pythagoreans, or for Aristotle. Pythagoras believed that he was the reincarnation of Euphorbus, a famous warrior, and "frequently sang the Homeric verses pertaining to himself, to the music of his lyre."[7]

One indicator that things were getting out of hand with Tycho came when he covered a wall of his study with a giant mural of himself. He began to view himself as a kind of successor to Ptolemy, which in his eyes at least made him better than royalty.[8] After King Frederick died, Tycho's relationship with the new king deteriorated when he refused to tend to Frederick's burial site, which was on his land. He closed down his laboratory in 1597 and moved first to Copenhagen, then to Prague to take a position as imperial mathematician to the Holy Roman emperor, Rudolph II.

Rudolph II was a reclusive and somewhat eccentric ruler who had a fascination with both science of the sort practised by Tycho and the occult. He also loved any kind of machine or gadget. When Tycho showed up in a carriage that featured an odometer, a novel device that signalled the distance travelled by ringing two small bells, he immediately got the job. He hired as his assistant a young astronomer by the name of Johannes Kepler.

THE THEORIST

Kepler's father was a mercenary soldier, his mother an innkeeper's daughter. Weak and sickly as a child, Kepler got his education in his grandmother's pub in Leonberg (now in Germany), then at a nearby seminary, then at the university of Tübingen, where he learned astronomy, mathematics, theology, and philosophy. His first job was at a new Lutheran high school in Graz, where he also served as district mathematician for the surrounding region. This meant preparing astrological calendars, along with forecasts of weather, crops, politics, and so on. In 1595, for example, he made three predictions: for a freezing cold winter, an attack by the Turks, and a peasant uprising.[9] Everyone was impressed, if a little inconvenienced, when all three came true. Six months on, he wrote to a friend: "By the way, so far the calendar's predictions are proving correct. There is an unheard-of cold in our land. In the Alpine farms people die of the cold. It is reliably reported that when they arrive home and blow their noses, the noses fall off. . . . As for the Turks, on January 1 they devastated the whole country from Vienna to Neustadt, setting everything on fire and carrying off men and plunder."[10]

Just as the Greeks had consulted with the oracles, the nobles and peasants of Europe ordered their world astrologically. The slow, ordered dance of the moon and the planets around the sky was a window into the mind of God, and a signal for success or disaster. Europe at the time was enveloped in the Little Ice Age, with unusually bad winters, a high risk of crop failure, and a population susceptible to the spread of epidemics (see boxed text on pages 68–69). Farmers and merchants used astrological clocks, like the one in Prague's Old Town Square, to aid their decision-making, and prognostication was a lucrative business. The arrival of a comet could even have implications for national security. Rudolph II employed scientists like Tycho and Kepler less for their scientific theories than for their usefulness at making astrological forecasts. He also hired savants such as the Englishmen Edward Kelley and John Dee, who obtained their visions of the future from crystal balls, mirrors, occult manuscripts, and angels.[11]

By the time he took the job with Tycho, Kepler already had his own ideas about the way the world worked. These were an intriguing mix of the old and the new. Rather than just try to produce a model that fitted the data, he wanted to know why things were the way they were. He was in search of a dynamic, a causative principle. Like Pythagoras, Kepler wanted to "perceive the true principles and causes of the Universe."[12] He believed that the universe was based on geometry, and that understanding geometry was the key to unlocking the workings of the universe. The Copernican heliocentric model had a simplicity and elegance, it seemed to Kepler, which was indicative of God's plan.

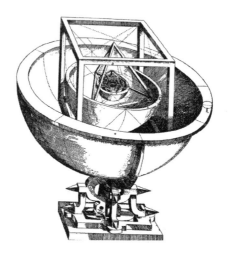

FIGURE 2.3. Kepler's model, based on Platonic solids.[13]

If Copernicus was right, his model could be used to estimate the relative sizes of the planetary orbits. It turned out that these were very different, so there were enormous distances between the planets. If this was part of God's plan, then what was the reason? Kepler was convinced that the answer lay in the Platonic solids, to which Pythagoras had attached great importance. Rather like Vitruvius, who stretched a man to fit a geometric pattern, Kepler believed that the relative sizes of the planets' orbits (only six were known at the time) could be accounted for by a nested sequence of spheres, each separated from the next by one of the Platonic solids (as shown in figure 2.3). Kepler published his theory, which the physicist B. K. Ridley describes as "pure Pythagoreanism," in his modestly titled *Mysterium Cosmographicum,* or *Mysteries of the Cosmos.*[14] Since Kepler's book supported Copernicus's heliocentric model, his position was unpopular, for religious and scientific reasons. To comprehensively prove his theory, he needed better data. Which is how he found himself applying for the job with Tycho.

SQUARE VS OBLONG

According to the scientific method first proposed by Aristotle and later developed by Sir Francis Bacon, scientists come to their conclusions by making careful observations, then forming a hypothesis, and then testing the hypothesis with experiments. If the theoretical predictions match reality, then the theory is confirmed, at least for the time being; otherwise, the theory must be changed. A type of conflict, between theory and experiment, is therefore built into the scientific process; but the aim is always to resolve it by coming up with a single model that satisfies these dual aspects.

To be a good observer requires patience, a mind for detail, the will to grapple with physical equipment. The ability to theorize often requires the opposite: a willingness to focus on abstract problems, a certain detachment from the everyday distractions of the real world. Plato told a story of the mathematician Thales, who one night was gazing at the sky as he walked and fell into a ditch. A servant girl helped him up and said, "How do you expect to understand what is going on up in the sky if you do not even see what is at your feet?"

If Tycho was the consummate observer, then Kepler was the ideal theorizer. The two had completely different personalities. The twenty-nine-year-old Kepler was a devout Lutheran who just wanted the peace and quiet to get on with his work, while Tycho, at fifty-three, was a rowdy, pugnacious, extroverted nobleman. His nose had been rebuilt from silver and gold after he'd had a fight as a student, and he maintained a dwarf named Jepp as court fool. But Kepler needed detailed observations if he was going to confirm his model of the universe. Even if he'd had the aptitude and the equipment, he had bad eyesight from an early bout with smallpox. Tycho had the observations, but he didn't have the theoretical ability to piece them together into a convincing model.

To Kepler, the problem with both the Greek Circle Model and Tycho's hybrid version was not that they didn't work, but that they were ugly. To match observations of the planets, they both had to introduce highly complex epicycles, circles within circles. A kind of corollary to the scientific method, known as Ockham's Razor, is that a theory should be no more complicated than necessary: that which is not needed should be cut away. It has since been restated in various forms. Newton wrote, "We are to admit no more causes of natural things than such as are both true and sufficient to explain their appearances." Einstein is supposed to have said, "Theories should be as simple as possible, but no simpler."

The desire for simplicity is married to a love of elegance, a kind of mathematical aesthetics. For Kepler, it was also a religious impulse. He believed that the motion of the planets reveals a kind of cosmic harmony. As Iamblichus wrote of Pythagoras, he "extended his ears and fixed his attention, his intellect, in the sublime symphonies of the world, hearing and understanding the universal harmony and consonance of the spheres and the stars that are moved through them, which produce a fuller and more intense melody than anything effected by mortal sounds."[15] God had given us the gift of mathematics so that we could understand and participate in this celestial music. So that we could join the dance.

Tycho and Kepler worked together for a year and a half. It was a battle between two scientific egos. Tycho was stingy with his data and released it to Kepler only in small doses, all the time trying to convert him to his hybrid model. Kepler, meanwhile, was trying to obtain enough data to verify his own model, and trying to get his salary improved so he could provide adequately for his wife and family. On October 24, 1601, the battle ended when Tycho passed away from complications after a urinary infection, which might have been induced when he tried to self-medicate with mercury (like many astronomers at the time, he was also a dabbler in medicine). On his deathbed, he pleaded to Kepler, "Let me not seem to have lived in vain."

Kepler inherited both Tycho's position as imperial mathematician and access to all of his data. Eventually, the data revealed to him that not only was Tycho's model wrong, but so was his own. At first, Kepler tried to adjust both to fit Tycho's observations. But after months of analysis, he realized that although planetary motion was nearly circular, it was better described by an ellipse, with the sun at one of the foci. The calculations were incredibly arduous; Kepler had at his disposal only primitive tools, such as basic geometry and trigonometry.

While Kepler is today best known for his discovery that the planets follow elliptical orbits, he seems to have taken far more delight in his earlier "discovery" that the universe is based on Platonic solids. In his notes to the second edition of *Mysteries of the Cosmos,* written twenty-five years later, ellipses are not even mentioned. As Arthur Koestler writes: "Kepler set out to prove that the solar system was built like a perfect crystal around the five divine solids, and discovered, to his chagrin, that it was dominated by lopsided and undistinguished curves; hence his unconscious taboo on the word 'ellipse.'"[16] He spent a full year trying to express the ellipse in terms of symmetrical circular motion, or as he put it, "squaring my oval."[17] A circle or square can be described by a single parameter, the diameter or diagonal, while an ellipse or rectangle seems arbitrary. To conclude that the universe was built on such figures was like suggesting to Brahe that he remodel his observatory garden after a randomly shaped oblong.

ODD VS EVEN

The culmination of Kepler's work was his book *Harmony of the World*. The title might have been inspired by a book by an Italian musician called Vincenzo Galilei, who believed that the music of the time could be revitalized by a return to the Pythagorean theory of harmony.[18] Kepler's book attempted to do the same for astrology, meteorology, and astronomy, by showing how they could be understood in terms of Pythagorean geometry. Pythagoras had claimed to be able to hear the celestial music of the planets as they orbited the earth; Kepler went further and figured out the notes. He said that "the Earth sings Mi-Fa-Mi, so we can gather even from this that *Mi*sery and *Fa*mine reign on our habitat."[19]

Buried in the quasi-mystical text was the last of Kepler's three laws of planetary motion. Together, these stated that

1. the planets rotate around the sun in elliptical orbits;
2. orbiting planets sweep out equal areas in equal time;
3. the squares of the period are proportional to cubes of average distance from the sun.

These rules held for all the planets. Like the theorem of Pythagoras, they were general principles that could be applied in different cases. Kepler saw them as manifestations of a force, the *anima movens* (which he equated with electromagnetism), that swept the planets around the sun. He was perhaps the first scientist to see the universe as a system with an underlying dynamic. A good theory does more than just reproduce data; it also gives a sense of why things happen. It tells a kind of story. The search for order is a search for meaning. Kepler's identification of a single, universal force was the precursor to Newton's law of gravity. Just as Renaissance painters such as Leonardo da Vinci had knocked the fixed, crystalline halos off religious art with a new sense of dynamism and movement, so Kepler broke away from the ancient concept of planets in crystalline spheres to the view that they were actively propelled by dynamical forces. "My aim," he wrote in a 1605 letter, "is to show that the machine of the universe is not similar to a divine animated being, but similar to a clock."[20]

The need to search for order and pattern seems to be a fundamental characteristic of human beings. Science, like religion, is a way of structuring and making sense of the world, a kind of bastion against chaos. While Kepler was working on books with titles like *Secret of the Universe* and *Harmony of the World,* his own world was in complete turmoil. Several of his children died from diseases that seemed to come from nowhere; the countries where he lived were embroiled in religious wars; his career was in the hands of a rather unreliable emperor; he was denied university positions because of his battles with religious orthodoxy; and finally, his mother, at around the age of seventy, was accused by her townsfolk of being a witch.

It is this last that shows the negative, shadow side of our desire for order. In a time when a child could die for no apparent reason, or a whole village could be devastated by plague, or a city could starve because of crop failure, it is not surprising that people would try to find scapegoats. Kepler's mother, like thousands of women across Europe, was one of them. She was apparently a strange, argumentative woman with an interest in folk medicine. We would probably call her odd, but in the Pythagorean scheme, which associated odd numbers with the divine and even numbers with conflict, her number would be even. A dispute with a neighbour grew, through gossip and innuendo, into the accusation of witchcraft. Specifically, she was charged with injuring seven people by giving them potions or hitting them, killing three more, and riding somebody's cow late at night (this was part of the folk tradition about witches). She spent over a year in prison while her case went to trial. It was only the intervention of her son—himself somewhat compromised by his disagreements with religious authorities—that saved her from being tortured and burned to death. Such a fate was not uncommon, and in fact had been suffered by the aunt who raised her.

It didn't play in Kepler's favour that as a young student, he had written a short piece of early science fiction that illustrated Copernican theory by imagining a flight to the moon to view the earth from an outside perspective. In the story, the boy is given the power to fly by his mother, who summons the spirits to usher him into the sky. This is the kind of thing you don't want read out at a witchcraft trial.[21]

About the only thing in Kepler's life that was predictable was the planets. And even they took on a different appearance when an Italian astronomer called Galileo Galilei, son of the musician whose book on harmony Kepler had read, took a custom-modified version of a new instrument known as a telescope and aimed it at the heavens.

THE LITTLE ICE AGE

The cold winter that Kepler correctly predicted in 1595 occurred in the depths of the Little Ice Age, a period of unusually chilly weather in Europe, North America, and elsewhere that started around 1550 and didn't end until the mid-nineteenth century. It is associated with the widespread advance of glaciers, the freezing of rivers such as the Thames, frequent famines, increased storms and floods, and even the decline of Viking colonies in Greenland and Iceland. The Little Ice Age is also reflected in the art of the time, like the winter landscapes of the Flemish painter Pieter Brueghel the Younger.

What caused the Little Ice Age? Some think the blame can be pinned on sunspots—intense electromagnetic disturbances that appear as darker areas on the surface of the sun. The solar radiation that we receive from the sun varies slightly in intensity; the lower the sunspot activity, the lower the total radiation. The Little Ice Age correlates with a period of reduced sunspot activity known as the Maunder minimum (though Galileo still managed to see a few). Sunspot activity also fluctuates over a period of about eleven years; the Victorian economist William Stanley Jevons believed this cycle was linked to agricultural production. While sunspot fluctuations are no longer thought to strongly influence the short-term weather, their electromagnetic pulses can disrupt satellite communications or even the power grid.

Many scientists believe that the Little Ice Age ended because of global warming; others argue that what we see as global warming is in part the natural ending of the Little Ice Age. In complex systems like the climate, the answers are rarely clear-cut.

AT REST VS IN MOTION

When science takes one of its occasional lurches forward, it is often because of some technological innovation. The magnifying glass had been used as an aid for reading since the 1200s, and by 1300 lensmakers were regularly grinding convex lenses for elderly, long-sighted people. These are easier to make than the concave lenses required for short-sightedness, which are thin in the middle and more easily broken. It was not until 1608 that a Dutchman called Lipperhey combined the two lenses to make a telescope.

Galileo learned of the new invention within months; realizing the instrument's potential, he set upon improving it. By learning to grind his own lenses and increasing the strength of the concave component, he boosted the magnification from 3X to 9X and more. He demonstrated his telescope to the Venetian senate and pointed out its obvious military applications. In a letter to the doge, he described how it enabled one "to discover at a much greater distance than usual the hulls and sails of the enemy, so that for two hours or more we can detect him before he detects us."

While the military applications of the telescope guaranteed Galileo's continued funding, the forty-six-year-old mathematics lecturer was soon pointing his lenses at the sky rather than at boats. Classical astronomers insisted that the moon was a perfect spherical orb, but Galileo was seeing what appeared to be craters and mountains. He also discovered four moons around Jupiter. Having realized that pure science could pay if done properly, he achieved another funding coup—and a position as mathematician and philosopher to the grand duke of Tuscany—by naming the moons after the wealthy Medici family. (They are now known as the Galilean satellites.) The fact that Jupiter had moons appeared to back up the Copernican theory, since it made it more plausible that our moon could go around the earth while we circle the sun. Furthermore, the ancient belief that the heavenly bodies were perfect spheres was undercut by the observations of features such as sunspots.

Galileo was already skeptical about the classical Greek Circle Model, not because of his observations of the heavens, but because of his far more prosaic experiments here on earth. One of Aristotle's theories had been that objects of different masses fall to the ground at different rates. While the effect of air resistance means that this might be true of a feather and a stone, Galileo showed that it was not generally the case. The story goes that he proved his point by dropping stones of different mass from the Leaning Tower of Pisa. He also demonstrated it by means of a thought experiment: if a falling brick broke in two in midair, then according to Aristotle the two pieces should slow, instead of continuing at the same rate. If the Greeks were wrong when it came to the motion of stones and bricks, then they could also be wrong when it came to the motion of planets.

The only way to divine the true nature of the universe, Galileo believed, was through careful observation and the use of mathematics. As he wrote in 1623, echoing both Pythagoras and Leonardo: "Philosophy is written in this grand book—the universe—which stands continuously open to our gaze. But the book cannot be understood unless one first learns to comprehend the language and interpret the characters in which it is written. It is written in the language of mathematics, and its characters are triangles, circles, and other geometrical figures, without which it is humanly impossible to understand a single word of it; without these one is wandering about in a dark labyrinth."[22] An object's qualities, like taste and smell, were secondary to the quantifiable properties of position, shape, number, and motion. A stone of a different colour would fall as fast.

In 1632, Galileo published his *Dialogue Concerning the Two Chief World Systems,* which argued for the Copernican model over the Greek Circle Model. (He ignored Kepler's version, still prefer-ring circular motion to ellipses.) It was presented as a discussion between three characters, called Salviati, Sagredo, and Simplicio. The simple-minded Simplicio represented the classical Greek view-point, and the character was widely assumed to be based on the pope of the time. Salviati represented Galileo's position. Sagredo pretends to be neutral but often backs Salviati, as here, where he mocks the dryness and sterility of the Aristotelian view of the cos-mos: "I cannot without great astonishment—I might say without great insult to my intelligence—hear it attributed as a prime perfec-tion and nobility of the natural and integral bodies of the universe that they are invariant, immutable, inalterable, etc., while on the other hand it is called a great imperfection to be alterable, gener-able, mutable, etc. For my part I consider the Earth very noble and admirable precisely because of the diverse alterations, changes, generations, etc., that occur in it incessantly. If, not being subject to any changes, it were a vast desert of sand or a mountain of jasper, or if at the time of the flood the waters which covered it had frozen, and it had remained an enormous globe of ice where nothing was ever born or ever altered or changed, I should deem it a useless lump in the universe, devoid of activity and, in a word, superfluous and essentially nonexistent. This is exactly the difference between a living animal and a dead one."[23]

Like Plato, Galileo had a flair for writing dialogue. He wrote some not terribly successful plays, and as a young man, he had even ventured into art–science fusion territory by delivering two lectures on the topography of Dante's *Inferno* based on a scientific analysis of the text. It was probably his taste for writing and drama that got him into trouble. If he had composed the usual dry, sterile scientific text, presenting the Copernican system as only a theoretical pos-sibility, then his work probably would have received little attention from the Inquisition. Instead, he wrote an entertaining dialogue in which the classical model was completely ridiculed. Most dangerous, it was in Italian, so it could be widely understood.

Priests of any religion like to have control over what gets written, and one of the best ways to do this is to use a specialized language. A century before Galileo, the Englishman William Tyndale was so upset by the foolishness and corruption of the local clergy that he decided to circumvent them and publish a version of the Latin Vulgate Bible in plain English. The Church strongly disapproved, since this would mean that priests would lose their monopoly as interpreters of the word of God. The archbishop of Canterbury had already declared translation of any part of the Bible a heresy punishable by burning. (And he wasn't talking about the books.) Tyndale was forced to leave England. In 1524 he travelled to Germany, then kept moving from place to place to evade the attention of English agents and assassins. All the time, he wrote and published. His enemy Thomas More complained that he "is nowhere and yet everywhere," like an early version of an Internet publisher. At the age of forty, in Antwerp, Tyndale was trapped by an agent and arrested. He was charged as a heretic and burned at the stake.

In Galileo's case, he was hauled in front of the Inquisition and forced to recant his view that the earth went round the sun. Unlike Tyndale, as well as the dissident scholar Giordano Bruno and many thousands of accused witches, Galileo managed to avoid the stake, but he was put under house arrest.

Being confined to his villa didn't stop him from working. In 1638, a copy of his *Discourse on Two New Sciences* was smuggled out and published in Leyden. The "two sciences" referred to the study of objects at rest—that is, mechanical properties of static objects—and in motion. Much of the book was based on experiments he had performed earlier but not yet written up. The work also represented Galileo's main contribution to the field of prediction, and explored the concepts of time, velocity, and acceleration.

These ideas seem mundane to any modern person who is used to working the accelerator and brake pedals in her car to achieve the right speed and get to her destination on time. But in Galileo's age, little was known about how things move. The problem with moving objects, after all, is that they don't stay still; a stone dropped from a tower was soon moving too fast to be timed using the primitive clocks of the day. Galileo realized, however, that he could slow down the experiments by studying the motion of balls rolling on an inclined plane, rather than in free fall.

He prepared a wooden plane with a channel about the width of a finger. He then rolled a bronze ball down the channel, starting from different heights, and timed its descent using a simple water clock. He reasoned that if the ball always fell at the same rate, then the descent time should vary directly with the initial height. So if the initial height was increased by a factor of four, then the amount of water released by the water clock during the fall should increase by the same factor. Instead, he found that the time increased only by a factor of two, and in general it varied with the square root of the height. It therefore followed that the object was speeding up as it fell down. He proposed that a falling object experiences uniform acceleration. If it is travelling at a given rate after one second of descent, then after two seconds it will have doubled its speed, and so on. He had succeeded in finding an unvarying principle, a kind of stillness at the heart of motion. The object moves, but the law that underpins the motion remains at rest.

Kepler showed, using the observations of Tycho, that the sun, earth, moon, and planets could be viewed as a single system that followed geometric laws and was united by a single force. Galileo studied the dynamics of motion here on earth and demonstrated that falling objects accelerate in a uniform fashion according to a mathematical law. It was Isaac Newton who synthesized these developments, proving that the plurality of falling objects here on earth and the motion of planets in the solar system were different instances of the same underlying principle: gravity.

ONE VS PLURALITY

Isaac Newton was born, prematurely, in Lincolnshire, England, on Christmas Day in 1642. The Christmas present might have been a little overwhelming for his mother, since her husband, an uneducated farmer, had died just three months before. Three years later, she married and went to live with a wealthy clergyman from the next village, leaving young Isaac behind to be looked after by his grandmother.

As a boy, Newton showed great talent at making models, a skill that proved useful when he began constructing his own experimental apparatus. He did well enough at grammar school, and was bad enough at farming, that it was decided he should attend Trinity College, Cambridge. He had to pay his way by waiting on faculty members and the richer students. His initial aim was a law degree, but he also studied mathematics and philosophy. In 1665, the university closed down when the plague, which had been sweeping its way across Europe, reached Cambridge.[24] Newton retreated to Lincolnshire, and started to work on problems of mathematics and physics.

The notoriously private and anti-social Newton later said that it was during these two years of near isolation that he formed his main theories. In perhaps another myth, a scientific version of the Garden of Eden, he observed the famous apple falling from the tree in his garden and, in a sudden flash of insight, realized that . . . it was lunchtime. Or that planets and apples are one and the same. Or that objects attract each other with a force that decreases in strength with the square of the distance. (The mathematician Karl Gauss thought the apple story, about the discovery of the law of gravity, was something Newton had fabricated to get rid of an annoying questioner.)

In 1667, once the plague had receded, Newton returned to Cambridge to continue his studies not just of mathematics and philosophy but of alchemy. While at grammar school, he had lodged with the local pharmacist, and he was fascinated by the smells and textures of the different chemicals that were used to make up the treatments of the time. The passion for experimenting with chemical compounds never left him, and he continued to pursue alchemy throughout his life. He also dabbled in other areas of mysticism, such as kabbalism, and searched for prophetic passages in the Bible. In fact, it could be said that Newton's scientific work was only a part-time pursuit, since his religious writings—including a 300,000-word tract on the book of Revelation that was published for the first time in 2004—constitute most of his output.[25] He believed that his scientific discoveries were in fact rediscoveries of ancient knowledge that had been passed from God to Noah to Moses to the Egyptian philosophers, and from them to Pythagoras (whom he credited with knowing the law of gravity).[26]

Like Galileo, Newton had his first public success with an improved telescope. The invention, which increased magnification by means of a mirror he ground himself, secured Newton a place in the Royal Society, a prestigious club for eminent scientists. There, he got drawn into a discussion with Edmond Halley (for whom the comet is named) and others about whether the elliptical planetary orbits that Kepler had discovered could be the result of an attractive force from the sun. Newton claimed that he had solved the question several years earlier, during his time in Lincolnshire, but couldn't find the proof. He wrote it up anew, starting a project that culminated in his seminal work, *Philosophiae Naturalis Principia Mathematica* (*Mathematical Principles of Natural Philosophy*), known more simply as the *Principia*.

The *Principia* laid down three laws of motion that Newton believed governed everything from apples to planets. The first law, often known as the law of inertia, stated (in Latin, of course) that "every body perseveres in its state of rest, or of uniform motion in a right line, unless it is compelled to change that state by forces impressed thereon." The key phrase here is "uniform motion." So if a person is hurtling across the ice on frictionless skates, he will continue moving in the same direction and at the same speed until he encounters an obstacle, such as an ice-hockey player.

Newton's second law states: "The alteration of motion is ever proportional to the motive force impressed; and is made in the direction of the right line in which that force is impressed." As Galileo knew, the speed of a falling object changes with a constant acceleration that is proportional to the force of gravity acting on it. The third law says: "To every action there is always opposed an equal and opposite reaction." The earth attracts the moon, but the moon also attracts the earth. The reason the moon goes around the earth, rather than vice versa, is because the earth is much larger. However, Newton showed (and Kepler had earlier argued) that the tides are the ocean's response to the moon's gravitational tug.

Scientists and military engineers, including Leonardo da Vinci, had long been interested in the behaviour of projectiles such as cannonballs. Newton carried out a kind of thought experiment in which he imagined a cannonball being shot horizontally from the top of a high mountain. If there were no air resistance, the ball would fall to the ground in a slow arc. If the cannon was extremely powerful, though, it could in principle send the ball over such a distance that the curve of the earth would have to be taken into account. There would come a point where the ball would continuously fall to earth but never get there, eventually performing a complete circle. In effect, this is what satellites, or the moon, do; they are always falling to earth but keep missing.

MALE VS FEMALE

To prove his results mathematically, Newton had to develop the tool of calculus, which was also independently discovered by Leibniz. Since this approach is key to how predictions are currently made in all branches of science, it is worth discussing in detail. Let's suppose we wanted to predict how long it would take a stone to hit the ground after being dropped, Galileo fashion, from the Leaning Tower of Pisa. The motion of the stone can be described at any time (t) by its position (x) and its velocity (v); these are known as variables, since they change with time. The laws of motion then give two equations:

$$dx/dt = v$$
$$dv/dt = -g$$

In words, the first line says that the rate of change of height x (denoted dx/dt) equals the velocity. The second line says that the rate of change of velocity (the acceleration) is given by the force of gravity per unit mass, which for objects near the earth's surface is a constant parameter g. Experiments have shown that g is about 9.8 in standard metric units (Newton showed it actually decreases with the square of the distance from the centre of the earth, but this is a small effect here). The minus sign indicates that the force is directed downwards, so height decreases with time.

The equations therefore involve two kinds of quantities: the variables x and v, which change with time, and the parameter g, which is treated as fixed. Time itself is treated as a dimension rather like distance, and is independent of the other quantities. As Newton wrote in the *Principia,* "Absolute, true, and mathematical time, of itself, and from its own nature flows equably without regard to anything external." It varies in a smooth, continuous fashion, and can be infinitely divided into smaller and smaller sections.

The two equations are known as ordinary differential equations, or ODEs, and together make up a dynamical system. Note that they tell us only how the position or the velocity is changing. They are similar to the prompts given by a navigation system in a car: drive straight now; turn left now; turn right now. If you follow the instructions at each moment, you should reach your destination even if you are in unfamiliar territory. The ODEs do the same thing, except that they constantly update the information. To predict when the stone will arrive at its destination, the ground, we have to compute all the forces acting on the stone at every instant.

The ODEs therefore tell a kind of story (or ode) about what happens to the stone from the moment it is dropped. If we probe a little deeper, we discover there is also a subtext. Francis Bacon wrote in *The Masculine Birth of Time* that scientific laws "conquer and subdue Nature," and they "storm and occupy her castles and strongholds."[27] In her more recent *Sideways Look at Time,* the author Jay Griffiths observes that "Aristotle used the terms 'male' and 'female' to describe differing understandings of time in the cosmos, calling the heavens male because he considered them eternal and immutable, while calling the earth female because it was changeable."[28] In this drama, then, the variables are the "feminine," changeable quantities that are governed and controlled by the "masculine," unchanging physical laws. The style is austere—the ODEs describe where the stone is and what it is doing, but they say nothing about its colour, shape, smell, texture, or weight. The senses are not engaged. As the Pythagoreans said, "'Tis mind that all things sees and hears. What else exists is deaf and blind."[29]

Calculating the solution to the ODEs—the actual path followed by the stone—is a thorny problem, but fortunately for us, Newton developed his theory of calculus to help. First, we must know the initial condition for the variables: the starting position x_0 (the height of the tower, which is about sixty metres) and the velocity v_0 (zero, if the stone is dropped). As shown in the notes, the answer can then be expressed in terms of x_0, v_0, t, and the fixed parameter g.[30] For any initial condition, the solution gives the height and the velocity as a function of time, so it can be viewed as a kind of machine: put in the time, turn the handle, and out comes the answer. For the solution to be accurate, the initial condition must be precisely measured and the equations must account for all the forces. Observations and theory must each be correct.

While we now have a formal answer to the problem, we still need a way to visualize and interpret the results. In his work *La Géométrie,* published in 1637, the French philosopher and scientist René Descartes welded together the Arab tradition of algebra, which dealt with equations, and the Greek tradition of geometry, which dealt with geometric figures, by showing that the solutions to equations could be plotted as figures on what is now known as a Cartesian grid.

THE GRID

Like many North Americans, I grew up in a Cartesian grid. The streets in Edmonton, Alberta, run north to south, while the avenues run east to west. The city is therefore divided into neat squares. This doesn't quite hold everywhere, of course—the river valley is stubbornly non-Cartesian, as are some of the suburbs—but on the whole, the system works, making it incredibly easy for Edmontonians to find their way around. In cities designed around more organic principles, I find it hard to shake the feeling of being permanently disoriented, like an explorer without his magnetic north.

Descartes reportedly got his idea when, as a bored young engineer working for the military, he contemplated a fly buzzing around the room and realized that it would be possible to represent its location in three-dimensional space by using three different co-ordinates.[31] (The idea later turned out to have all sorts of military applications, in guidance systems, for example.) If the aim is instead to locate an office in Edmonton, the co-ordinates could be the street and building number, the avenue, and the floor number. In a dynamical system, the method can be used to plot any variable or combination of variables. Time can be treated as a separate dimension, like a spatial co-ordinate.

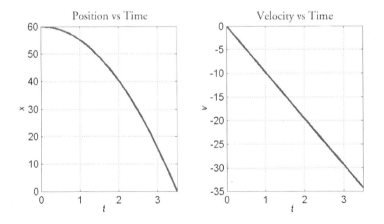

FIGURE 2.4. Plots of the position (x) and velocity (v) of a falling object, as a function of time (t).

Figure 2.4 shows x and v as functions of time. If you look up a certain time on the horizontal axis of the Cartesian grid, you can find the value of the variable at that time on the vertical axis. In the left grid, the stone begins at time 0 at a height of 60 and takes about 3.5 seconds to reach height 0 (i.e., hit the ground). The right grid shows that velocity becomes increasingly negative (i.e., faster) in a linear fashion.[32] In reality, the speed would taper off because of the effect of air resistance, which is not considered here.

If the equations of the dynamical system are sufficiently simple, then the methods of calculus can be applied to find the solution, as here. More complicated systems often cannot be solved using calculus, so an approximation technique must be employed. Essentially, this involves dividing the time into short steps, solving the equations for each time step, and finally stitching the answers together. This approach involves a lot of computation and can be numerically unstable, so it didn't become feasible for general use until the invention of fast computers in the 1950s.

Newton was a Unitarian and rejected the doctrine of the Holy Trinity. (In religion as well as science, he favoured one over plurality.) If this had become known by his Cambridge employers, it would have cost him his job. But unlike Kepler and Galileo, Newton kept his views secret and avoided such fights, though the stress might have contributed to an apparent nervous breakdown he suffered in 1693. He cut back on scientific research and took a position as master of the mint at the Bank of England. (Pythagoras was not the last scientist to take an interest in coins.) He took his job seriously, especially when it came to combatting counterfeiters, several of whom he sent off to the gallows. He also made sure to protect his own image, fiercely denouncing scientists such as Robert Hooke who could claim to have contributed to his ideas. He died in 1727 a wealthy and famous man, and was buried in Westminster Abbey.

Alexander Pope supplied his epitaph, which compared Newton to a kind of cosmic light bulb:

> Nature, and Nature's Laws lay hid in Night.
> God said, Let Newton be! And All was Light.

Newton synthesized the work of many scientists before him and created a kind of Unitarian theory of the universe, which viewed everything from apples to planets as being united under a single set of laws. Kepler had discovered that the planets move in elliptical orbits, but he had seen this as a violation of Pythagorean symmetry. Newton showed that what counted was not the shape of the motion but the underlying dynamic. The orbit might be an ellipse, but the law is simple and square.

A consequence was that scientists suddenly seemed to have enormous predictive power at their disposal. If the motion of a planet around a star is given by a simple equation, then we can figure out its position at any time in the future just by plugging in numbers. And if phenomena here on earth obey similar principles, they too can be modelled mathematically. Kepler, it seemed, was right: we lived in a deterministic, predictable, clock-like universe.

3 ► DIVIDE AND CONQUER
THE GOSPEL OF DETERMINISTIC SCIENCE

The simplicity of nature is not to be measured by that of our conceptions. Infinitely varied in its effects, nature is simple only in its causes, and its economy consists in producing a great number of phenomena, often very complicated, by means of a small number of general laws.

—Pierre Simon Laplace, French mathematician and physicist

The vanity of men
they would like to retain
this passing winter moon.

—Issa, Japanese haiku poet

STRAIGHT VS CROOKED

I once worked for two years at a research institute in Saclay, near Paris. While living there, I was struck, among other things, by the enjoyment the French seemed to take in straight lines. The gardens of Paris and Versailles glory in the geometric perfection of the line that seems to go on forever. From the Louvre, you can look in an almost perfect line through the Jardin des Tuileries, down the Champs Élysées, all the way to the Arc de Triomphe and beyond.

Even the bread is in the shape of a line. Perhaps it is not surprising that so much deterministic science, based on linear, cause-and-effect thinking, made its start there. The apple of deterministic science may have fallen from Newton's tree, but it found particularly fertile soil in the country across the Channel.

In 1637, Descartes had prepared the ground with his *Discours de la méthode,* in which he laid out his principles for scientific discovery. These were:

- Never accept anything as true unless it clearly is.
- Divide every difficult problem into small parts, and solve the problem by attacking these parts.
- Always proceed from the simple to the complex, looking for patterns and order.
- Be as complete and thorough as possible, so that nothing is missed.[1]

That troublesome beast nature was to be interrogated like a witness in a court of law. To analyze a complex event, it was only necessary to break it down into separate components, work out where everything was at any given time and what it was doing, and apply the basic physical laws. Just as Baron Haussmann would soon be blasting arrow-straight boulevards down the centre of Paris, the French scientists, along with their colleagues across Europe, began applying the gospel of deterministic science to the natural world.

Newton had shown that the force of gravity varied inversely with the square of distance, and he'd used calculus to demonstrate that as a result the planets followed elliptical orbits. Charles Augustin de Coulomb then found a corresponding inverse-square law for electrical charges. If the distance between two charge-carrying spheres is doubled, he determined, the force between them decreases by a factor of four, exactly like gravity. It followed that charged bodies were as predictable as the planets.

Antoine Lavoisier pushed the Newtonian scheme into the field of chemistry. Newton had throughout his life been fascinated by alchemy, but he'd never examined the behaviour of chemicals with the same kind of rigour that he applied to the heavens. Lavoisier, whose day job as an accountant for a private tax-collection agency resulted in a date with the guillotine during the French Revolution, applied his accounting skills to chemical experiments. He showed that chemicals were of two types, elements and compounds. In a famous two-day public experiment, he demonstrated how water was made up of hydrogen and oxygen. We are now somewhat inured to the wonder that this must have caused the audience, but it is still amazing to think that water, perhaps the most important substance in our lives, is actually made up of two gases, oxygen and hydrogen, the second of which is highly explosive.

By the nineteenth century, science was on a roll—or rather, a relentless Napoleonic march down a long, straight line. The optimism was captured by Pierre Simon Laplace, who stated in 1820 that if we knew the present state of all particles of the universe, and the forces acting on them, we could, in principle, predict their future: "We may regard the present state of the universe as the effect of its past and the cause of its future. An intellect which at any given moment knew all of the forces that animate nature and the mutual positions of the beings that compose it, if this intellect were vast enough to submit the data to analysis, could condense into a single formula the movement of the greatest bodies of the universe and that of the lightest atom; for such an intellect nothing could be uncertain and the future just like the past would be present before its eyes."[2] This know-it-all intellect is sometimes referred to as Laplace's demon. Of course, Laplace didn't think scientists could achieve such accuracy in practice—but their goal was to come as close as possible.

Not everyone was convinced by the power of determinism, or by the idea that what worked for planets would apply everywhere. In 1790, Immanuel Kant said: "It is absurd to hope that another Newton will arise in the future who will make comprehensible to us the production of a blade of grass according to natural laws." Leaders of the Romantic movement, like Blake in England and Goethe in Germany, criticized the mechanical model and espoused an organic view of the world. Goethe's epic poem *Faust* turned the idea of selling the soul to the devil into a metaphor for the power and the hidden, human cost of materialistic science. Blake put it even more bluntly: "May God us keep / From single vision and Newton's sleep!"[3]

However, deterministic science was stunningly confirmed, at least to its practitioners, by its ability to predict previously unknown features of physical systems. James Clerk Maxwell produced four equations that showed that light was a combination of electric and magnetic oscillations, and that predicted the existence of electromagnetic waves outside the visible spectrum (a hypothesis that was later confirmed by Heinrich Hertz, and nowadays by anyone who listens to his car radio). Dmitri Mendeleyev predicted the existence of undiscovered substances to fill the gaps in his periodic table, which ordered the elements according to atomic weight. Urbain Leverrier used Newton's laws to show that measured perturbations in planetary orbits had to be caused by the presence of another planet, Neptune, whose existence was soon confirmed by observation. (He also predicted the planet Vulcan, which too was observed for some time, despite the fact that it didn't exist.)

Determinism wasn't limited to purely physical systems. In December of 1831, a twenty-two-year-old med-school dropout boarded the *Beagle* to begin a five-year trip around the world. At the time, many people relied on physiognomy to divine human character, and the captain of the *Beagle,* Robert FitzRoy, almost didn't select the young man because he distrusted the shape of his nose.[4] However, Charles Darwin passed this hurdle, by a nose, and started a voyage that would change both his life and our understanding of life.

LIMITED VS UNLIMITED

The object of the *Beagle's* expedition was "to complete the survey of Patagonia and Tierra del Fuego . . . to survey the shores of Chile, Peru, and of some islands in the Pacific—and to carry a chain of chronometrical measurements round the world."[5] During the five-year cruise, Darwin made copious observations of geology, plants, insects, animals, fossils, and the local human beings. He also produced some great travel writing: "In the morning we got under way, and stood out of the splendid harbour of Rio de Janeiro. . . . We saw nothing particular, excepting on one day a great shoal of porpoises, many hundreds in number. The whole sea was in places furrowed by them; and a most extraordinary spectacle was presented, as hundreds, proceeding together by jumps, in which their whole bodies were exposed, thus cut the water. . . . One dark night we were surrounded by numerous seals and penguins, which made such strange noises that the officer on watch reported he could hear the cattle bellowing on shore. On a second night we witnessed a splendid scene of natural fireworks; the mast-head and yard-arm-ends shone with St. Elmo's light; and the form of the vane could almost be traced, as if it had been rubbed with phosphorous. The sea was so highly luminous, that the tracks of the penguins were marked by a fiery wake, and the darkness of the sky was momentarily illuminated by most vivid lightning."[6]

Back in the less distracting surroundings of his home in England, Darwin evolved the theory of evolution by natural selection, which was based on the idea that variety plus selection equals evolution:

Variety	+	Selection	=	Evolution
Offspring are similar, but not identical to, their parents. Each individual therefore has different qualities.		Every species produces far more offspring than can survive. Those that are better suited to their environment are more likely to live to have children of their own.		As a result, the species gradually evolves and becomes better suited to its environment.

Darwin's theory implied that we—like seals, penguins, porpoises, and all other creatures on the planet—are merely the end product of a long line of gradual, mechanistic changes and improvements.

That life evolves in some way had been proposed by other scientists, including Jean Baptiste de Lamarck, who first coined the word "biology" and believed that species developed directly in response to their surroundings. A proto-giraffe straining to reach the upper leaves might not make it himself, but he would pass on to his offspring a longer neck. Darwin's mechanism differed in that the variability was assumed to be purely random. He believed the environment acted as a selection mechanism, and therefore gave a direction to evolution, so that only favourable changes were retained over generations. (The theory didn't say exactly what caused the variation, but it was later traced to the genes.)

Darwin's theory was strongly influenced by the work of two economists. The first was Rev. Thomas Malthus, who in his 1798 *Essay on the Principle of Population* made a notorious prediction: that the human race would eventually outstrip its food supply. His essay didn't prove its points with equations, but it nevertheless presented a kind of mathematical model: "Population, when unchecked, increases in a geometric [i.e., exponential] ratio. Subsistence increases only in an arithmetical [i.e., linear] ratio. A slight acquaintance with numbers will show the immensity of the first power in comparison with the second." Suppose, for example, that every twenty-five years, on average, a couple produces four children who survive long enough to have children themselves. Since the replacement rate needed to maintain a stable population is only two people per couple, this means that every twenty-five years, the population will increase by a factor of two (see figure 3.1). We would say that the population grows in an exponential fashion (known to Malthus as geometric), with a doubling time of twenty-five years.

Sustenance, on the other hand, relies on increasing the harvest, which Malthus argued would increase only at a linear rate. Since exponential growth always overtakes linear growth, no matter how much land or how many people one starts with, the population will eventually become too large to support itself. Dire consequences follow: "Sickly seasons, epidemics, pestilence, and plague, advance in terrific array, and sweep off their thousands and ten thousands. Should success be still incomplete, gigantic inevitable famine stalks in the rear, and with one mighty blow levels the population with the food of the world."

Malthus's somewhat pessimistic conclusion was out of step with the expansionist mood of Victorian Britain, where large families were positively encouraged. (Darwin, for example, had ten children, though three died in childhood.) It helped lead to economics being labelled the dismal science, and turned him, in the words of one biographer, into "the best abused man of his age. . . . For thirty years it rained refutations."[7] Karl Marx, himself a determined determinist, called Malthus's essay a "libel on the human race."[8] Darwin, however, realized that this imbalance between the production of offspring and the resources of the environment would provide a kind of mathematical dynamic for his theory of evolution. With such intense competition for resources, only those best suited to the environment would survive and pass their attributes on to the next generation, while the weak would perish.

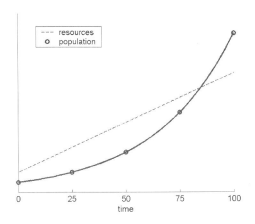

Figure 3.1. If population grows exponentially but resources linearly, then population will eventually outstrip resources, leading to famine.

The exponential growth in population identified by Malthus is a result, to use a modern term, of positive feedback; the rate of increase itself increases with time. It's like a bank account where the interest earned each year is reinvested; the amount of money available to earn interest the next year therefore increases, leading to the exponential growth that bankers call the "miracle of compound interest" (unless the account earns only a fraction of a percent a year, in which case it is the "miracle of high bank profits"). With a growing population, there are always more people to have babies, so the total rate of births increases, and so on. Growth begets growth.

In fact, most species have a much faster turnaround than humans, and some can grow at truly explosive rates. While the bacteria *E. coli* were unknown to Darwin—he was not unknown to them, since they resided in his gut—the species can replicate itself at a rate of about one generation every twenty minutes. In twelve hours a single cell, given ideal growth conditions, could grow to 2^{36}, or 68 billion cells. In less than one day, the colony could grow large enough to fill a room.

As Malthus pointed out, there is a regulatory mechanism that will eventually limit and control this growth. *E. coli* doesn't take over the world because as the colony increases in size, it runs out of food. The individual members come into competition, both with each other and with other species, for the available resources. The higher the population, the fiercer the competition, so continued growth makes future growth less likely—negative feedback. The population does not therefore continue exponentially, but instead tapers off at some limit. If it exceeds that level, the force of competition will restore it to equilibrium.

An example of such regulation in a different context was James Watt's invention of the flyball governor (see page 93). Darwin, an enthusiastic trader of company shares, was probably more influenced by the work of a second economist, Adam Smith. In *The Wealth of Nations,* Smith identified a regulating mechanism in free markets, which he referred to as the "invisible hand" of capitalism. If a shortage of some product occurred, for example, the price would spike up; however, as profits rose, more suppliers would enter the market, thus driving the price down once again. Conversely, if prices were too low, suppliers would leave or go broke. Prices would always therefore be restored to the "natural price," which would reflect the cost of production, as an automatic result of people's self-interest: "It is not from the benevolence of the butcher, the brewer, or the baker that we expect our dinner, but from their regard to their own interest. We address ourselves, not to their humanity but to their self-love." While positive feedback accentuates changes, be they increases or decreases, negative feedback acts as a stabilizer that damps out perturbations.

The Flyball Governor

A mechanical example of negative feedback is James Watt's 1788 invention, the flyball governor, which was used to control the speed of a steam engine. The device consists of two heavy metal balls suspended by movable arms from a central revolving spindle, which is driven by the engine. As the engine speeds up, the two metal balls spread apart and lift under the centrifugal force. They are linked to the throttle in such a way that if the engine runs too quickly, the power cuts back. Conversely, if the speed is too slow, the balls sink down, making the throttle accelerate. It's like a hand on the tiller of a boat, keeping a steady course.

The apparently simple mechanism results in quite complicated non-linear dynamics—perhaps ironically, systems designed for stable behaviour can be very hard to model. The physicist James Clerk Maxwell had to invent a new branch of mathematics, known as control theory, to analyze Watt's invention. Without the governor, steam engines would have blown up, so control systems were as important to the Industrial Revolution as coal. Positive and negative feedbacks are ubiquitous not just in engineering, but also in atmospheric, biological, and economic systems, which is one of the reasons predicting them is difficult.

Darwin viewed the natural world as a kind of self-regulating dynamical system of opposing forces. Positive and negative feedback were two manifestations of an underlying creative principle. Variation among individuals and survival of the fittest created a trial-and-error selection mechanism, which meant that species' traits did not remain static but were in a constant state of dynamic flux and self-improvement. Malthus saw a system spiralling off into disaster, but the Darwinian world was limited, kept in check, and constantly improved by competition. Just as a healthy economy was maintained by the invisible hand of price competition, nature was a well-tended garden, with the law of selection doing the pruning and weeding.

In one sense, then, Darwinism was Malthus with a happy ending. In another sense, though, the theory of evolution was as radical as Copernicus's claim that the earth went around the sun. The heliocentric model had dealt the human ego a huge blow, since it meant that rather than occupying a favoured position at the centre of the universe, our planet was just rotating through space with the others. Evolution implied that there was also nothing particularly special about the human race. We were evolving as well as revolving. Like Copernicus, Darwin was aware of the upsetting consequences of his work, and perhaps as a result, he didn't publish *On the Origin of Species* until 1859, twenty-eight years after his voyage, and then only because the naturalist Alfred Russel Wallace had come up with the same idea. One compensation of the theory was that it was widely agreed that English gentlemen, such as Darwin and his peers, were themselves the pinnacle of evolution.

THE ARROW OF TIME

By the end of the nineteenth century, science had itself evolved into a grand world-view, a kind of secular faith that rivalled, or for increasing numbers of people replaced, religion. Nature—for so long a source of awe and mystery, but also of fear—had been placed in the dock, interrogated, broken down. The laws had been applied. In Darwin's theory of evolution, science even had its own creation myth: nature had been shown to be the result of her upbringing. It followed that she could be predicted and controlled.

Newton had believed that the universe was rather like a game of pool. As he wrote in the *Principia*, "It seems probable to me that God in the beginning formed matter in solid, massy, hard, impenetrable, movable particles." Even light, he believed, was a stream of such particles (the colour balls). God was relegated to the role of prime mover, the one who breaks the pack with a brilliant cue shot and lets it go; after that, everything worked on automatic. But if all natural phenomena could be explained by physical laws, even the evolution of our own species, then there seemed to be little need for a divine force at all. When Napoleon asked Laplace why God did not play a role in his calculations for the solar system, he is said to have replied, "I have no need of that hypothesis."

This clockwork, materialistic view of the world applied also to the insides of living beings. Diseases that had once been viewed as random events, or the result of witchcraft, were shown by biologists such as Louis Pasteur to be caused by micro-organisms. Our own bodies were structured like a complex machine: the heart was a pump, veins and arteries the plumbing, the nervous system a network of electrical cables. Newton's doctor, Richard Mead, believed that one day medicine would reduce to a set of equations.[9]

Science also extended its cold, mechanical arm into the humanities. The fields of economics, political science, and sociology developed into quasi-deterministic sciences that mimicked the equations of physics in their modelling of human behaviour. Ensconced in the reading room at the British Library, Karl Marx developed his theory of class struggle, which would culminate, he believed, in the ideal state of communism—a kind of heaven for true believers. At Marx's burial service, his collaborator, Friedrich Engels, claimed, "Just as Darwin discovered the law of evolution in organic nature, so Marx discovered the law of evolution in human history."[10] Institutes such as the London School of Economics modelled society and the economy as Darwin-like competitions for limited resources. August Comte introduced the term "sociology" to describe the scientific study of social behaviour, but he preferred the name "social physics."

Even the workings of the mind could be explained in machine-like terms. Sigmund Freud shocked the Victorians in 1900 with *The Interpretation of Dreams.* The human mind, he suggested, was a complicated system of drives and forces, the most powerful of which was the sexual urge. If repressed, it sought release in dreams or, sometimes, in deviant behaviour. Like a blocked drainpipe in the plumbing of the soul, every psychological problem had a precise cause that could be resolved by calling in the analyst.

The revolution in science marched apace with the Industrial Revolution, which spread both scientific technology and Newton's concept of "absolute time" into factories and sweatshops around the world. Once, time had been a somewhat elastic, non-linear concept. Days were broken down into regular intervals, but these would be shorter in winter and longer in summer, to reflect the changing length of the day.[11] While Newton was building his mathematical theories, however, the Royal Greenwich Observatory was going up outside London. Its construction, funded by the sale of 690 barrels of gunpowder, was to ensure British naval supremacy, which relied on accurate navigation, which in turn required accurate clocks.[12] In 1884, the Greenwich Meridian became the world's prime meridian, dividing the globe, by a straight line, into East and West, so time and space were both on the same Cartesian grid. Uniform, linear time spread around the world, along with language and empire. Soon everyone was marching to the same Greenwich Mean beat. Time was parsed to smaller and smaller intervals, and people's lives became dominated by the absolute law of the clock and the worship of the twin gods of time and money (though Benjamin Franklin argued that time was money).

While the power attributed to nature or divine forces was on the decline, the power of the scientist was on the rise. Books such as *History of the Conflict between Religion and Science* (1874), by the New York University scientist John Draper, and *A History of the Warfare between Science and Theology in Christendom* (1896), by Andrew Dickson White from Cornell University, directly challenged the role of religion. To many people, science now seemed to hold the answers to the big questions: Who are we? Where do we come from? And especially, where are we going? The future of the world was not determined by random forces or destiny or the will of God, but was the natural consequence of physical laws. In the mechanistic church, the bible was the laws of physics as encoded in the scientific literature; the priests who interpreted these laws were the scientists; and the cathedrals were the laboratories, universities, and new research institutes like the Massachusetts Institute of Technology. The linear march of science seemed unstoppable—at least until, at the dawn of the twentieth century, it hit the first in a sequence of bumps in the road.

CHAOS

The success of reductive science lay in its ability to "divide every difficult problem into small parts, and to solve the problem by attacking these parts." For example, the methods of calculus, when used to predict the path of a falling stone, involve breaking down time and space into smaller and smaller increments. This works because time is, for the purposes here, uniform. One instant is like another—it's all the same stuff. Greenwich Mean Time is unaffected by the fact that someone is dropping a stone from a tower. In the case of time, it seemed true that one could divide every problem into small parts.

The divide-and-conquer approach also works for combinations of simple forces. If the falling object in figure 2.4 (see page 81) also moves horizontally—say, it has been shot from a cannon—then the velocity can be parsed into horizontal and vertical components, and each treated separately. In general, the systems that yield most easily to this approach are those that are described as linear. While the term "linear" is often used to describe the straightforward, cause-and-effect logic championed by Aristotle, it also has a more precise mathematical meaning. A function of some variable is linear if the plot of the function versus the variable is a straight line.[13] An example is a perfect spring, which is defined as one for which the restorative force varies linearly with displacement. Place a small weight such as a piece of straw on the spring, and it will compress a certain amount. Double the weight, and it compresses the same amount more. The initial condition of the spring doesn't matter—it changes the same extra amount whether there is a weight already there or not. To know the deflection caused by two or more straws, we can measure the deflection of a single straw and multiply by the total number. In a linear system like this, the whole is the same as the sum of the parts.

Now, speaking metaphorically, replace the spring with a camel. As you add pieces of straw onto the camel, at first little happens—the camel gradually sags lower and curses you in camel language. But suddenly you come to the "straw that breaks the camel's back," and you get what mathematicians term a "non-linear response." The impact of the last straw is much greater than the impact of the first. With non-linear systems, initial conditions matter. This fact was discovered in a rather unsettling fashion by Henri Poincaré, a professor at the University of Paris.

In 1889, to commemorate the sixtieth birthday of King Oscar II of Sweden and Norway, a competition was held to reward the best research in celestial mechanics related to the stability of the solar system. While we all know that the sun rises in the east and sets in the west, it isn't entirely obvious that this is a permanent state of affairs. Who is to say that the earth couldn't take a detour one morning and head off into outer space, with the sun just becoming a gradually shrinking dot? This was the kind of nagging question to which aging monarchs turned their attention.

The competition was won by Poincaré, who showed that there was no straightforward solution to a system with even three bodies. A small difference in the initial positions could lead to radically different outcomes. The reason for this mess was non-linearities in the equations. As if involved in a complicated, triangular love affair, each planet was affected in a non-linear way by interaction with the other bodies. Depending on the initial configuration, a planet could either stay in the system or be ejected from it. Furthermore, the trajectories were so complicated that Poincaré couldn't even draw them. He had discovered chaos.

When I was a kid, my parents gave me a pendulum, a rod with a magnet on its end, which swung just above a metal plate. There were three additional small magnets that could be positioned anywhere on the plate. When the pendulum was released, it would swing in a wild and unpredictable fashion as it was attracted to or repelled by the magnets, until it eventually came to rest because of friction in the rod. Poincaré's system was similar, except that instead of competing magnetic fields, there were gravitational fields and no friction.

Since the earth's orbit is constantly perturbed by the other planets, Poincaré's result implied that the solar system might be less

stable than it appears. Astronomers later confirmed that the solar
system is probably weakly chaotic, but the perturbations from regular orbits are small, so we won't crash into Venus any time soon.

Non-linear systems are not unusual. Just as the irrational numbers outnumber, so to speak, the rationals, so most systems are non-
linear, rather than linear. There are more ways to draw a crooked
line than a straight one. Chaos is a somewhat rarer phenomenon,
but it's still easily produced. As a simple example, imagine that you
take a lump of bread dough, stretch it to twice its length, then cut
it in half and put one piece on top of the other. Then you repeat, as
shown in figure 3.2. The procedure is similar to the usual kneading,
although there you would fold the dough back onto itself without
cutting it in half.

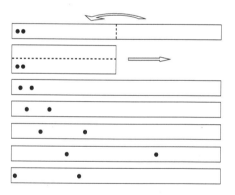

FIGURE 3.2. Two yeast cells in the dough are initially separated by
a distance *d*. The dough is cut in half, the two halves are stacked
together, and then the dough is stretched so the distance between
cells doubles to 2*d*. After another iteration, the distance is 4*d,* and so
on. This continues until one or both cells cross to the right-hand side
of the dough.

Consider two yeast cells in the dough, perhaps a mother and a daughter (yeast reproduces by budding). Each time the dough is stretched, the distance between the mother and the daughter increases by a factor of two. The system therefore shows sensitivity to initial condition: the mother and the daughter started at almost the same position, but after only a few iterations, they are tragically and irreversibly separated.

The system is known as the shift map, for reasons discussed in Appendix I. Like an ODE, a map takes you from a starting point into the future, but it does so in a discrete, step-wise fashion. In fact, when ODEs are solved numerically, as a sequence of steps of length Δt, they are equivalent to maps. Chaotic ODEs show both the sensitivity to initial condition and even the same kind of mixing or kneading behaviour as the shift map.

To see the effect that chaos has on predictability, suppose we try to predict the future location of the daughter cell based on the position of the mother cell. This small error in the initial condition will be magnified exponentially until the prediction becomes useless. Even with only a small uncertainty in the initial condition, any prediction performed by applying the shift map is soon no better than random. In fact, as shown in Appendix I, a better guess is to use as our "model" the long-term average of all possible states (i.e., we look for the yeast cell in the middle of the dough). This is sometimes referred to as the climatological forecast: it is like predicting the temperature for a day next week by dispensing with a dynamical model altogether and just determining the average from historical records for that particular day of the year. This illustrates how, even for a simple system, prediction strategies can be complicated by the effects of chaos.

The word "chaos" was actually first used to describe such systems by the mathematician James Yorke in 1975.[14] For the ancient Greeks, chaos was the original formless void from which the cosmos was born. Pythagoras associated it with the unlimited, indeterminate aspect of the universe, while the Roman writer Ovid described it as an unordered primordial mass. It therefore evokes a somewhat frightening image of complete randomness and emptiness of meaning, and indeed its existence threw a spanner into the nineteenth century's clock-like universe. Chaos implied that having an accurate model of a system wasn't enough to predict its future evolution. Even the slightest error in our knowledge of the initial state would eventually grow. And in a non-linear system, the strategy of breaking a whole down into its components would no longer apply because, like a haiku poem, the whole would be more than the sum of the parts.

Once chaos was discovered, though, it was mostly ignored. Powerful techniques existed to solve linear problems, but non-linear equations were hard or impossible to tackle, so scientists and engineers simply avoided them, at least until the invention of fast computers in the 1960s. Many phenomena could still be reasonably approximated by the linear approach, particularly engineered systems (which were often designed specifically to work in a linear regime). A metal beam with only small displacements was almost linear, as was the flow of water around a surface at slow speed. While chaos made the job of Laplace's demon harder, it did not make it impossible, nor did it call into question the mechanistic view of the universe. Even if there was uncertainty in the initial condition, it would still be possible to make probabilistic predictions by taking this variability into account.

HARD AND SOFT

Just as the mechanistic philosophy had managed to bypass chaos, though, and seemed to be taking over, robot-like, the entire world, it stumbled into another small obstacle, which caused it to pause. This was the awkward development, from two extraordinary 1905 papers by Albert Einstein, of the theories of relativity and quantum mechanics. These changed physics—the first at very high speeds, the second at very small scales. The world machine was not as deterministic as it seemed.

Galileo once argued that so long as you travel smoothly at a constant rate, it is impossible to tell whether you are moving. Suppose you are on a boat that sets off so smoothly that you don't notice its motion. If you look out the window and see a docked boat passing by, you might mistakenly think that it is moving and you are not. And so long as you are moving at a constant speed, you'll find that if you drop your book to the floor, it will fall straight down, exactly as it would if you were outside on the dock. The laws of physics apply whether you are moving or not. But according to the equations of James Clerk Maxwell, the speed of light should also be constant. Therefore, if a person on the dock flashes a light at you, the beam will appear to both of you to be travelling at the same speed, even though you are moving relative to each other. But this doesn't make sense—unless, as Einstein argued in 1905 with his special theory of relativity, time and motion are somehow linked. He predicted that time should flow at a slightly different rate depending on whether you sit still or whiz around very fast. (Scientists tested this in 1971 by comparing two highly accurate clocks, one on the ground and the other on an aircraft. The time difference recorded after a long flight, measured in millionths of seconds, was consistent with predictions made using Einstein's equations.) His theory was later extended to the general theory of relativity, which showed that gravity was a kind of distortion in the space-time fabric. Newton's reductionist law of gravity was thus reduced even further, to a by-product of relativity.

As an encore, Einstein suggested that a number of paradoxes associated with the classical theory of electromagnetic radiation could be resolved by assuming that it existed in discrete packets, or quanta. The full implications of quantum mechanics, as it became known, occupied the best physics minds of a generation, but the eventual conclusion was that matter and energy are interrelated (by Einstein's famous equation, $E = mc^2$) and have a dual particle/wave nature. In Newtonian dynamics, the state of a particle could be exactly described in state space by a single position and momentum. But in quantum dynamics, this was replaced by a wave function, which only specified the probability that the particle could exist in a particular region of state space. Furthermore, Heisenberg's uncertainty principle stated that the more accurately the position was measured, the greater the uncertainty in momentum, and vice versa. Matter was both determinate and indeterminate, limited and unlimited, hard and soft. At a fundamental level, nature did not observe such distinctions. The wave model and the particle model were incomplete: like breathing in and breathing out, they were two aspects of a unified, dynamic process.

Together, relativity theory and quantum mechanics unravelled much of the certainty that underlay classical physics. Time was not absolute and independent, but was itself a dynamic quantity that depended on the observer. The uncertainty principle showed that even Laplace's demon could never know the exact position and momentum of a particle. It was as if the universe had a built-in random component. If it wasn't possible to know the initial conditions of a system, and if time itself depended on the observer, then how could someone predict the future? Properties such as non-locality, where particles can influence each other instantaneously across space, seemed to undermine even the idea of cause and effect.

Such implications were fiercely resisted by some scientists. Einstein, who was religious, insisted that "God does not play dice with the universe," and pursued until his death the Pythagorean dream of a "theory of everything" that would reduce the universe to a small set of equations. He also spearheaded a rapprochement between science and religion, like a high priest capable of interpreting the mind of God. As he once remarked, "In this materialistic age of ours the serious scientific workers are the only profoundly religious people."[15] Others embraced quantum physics as the way to a new science that equated matter with a kind of spirit (see, for example, Fritjof Capra's *Tao of Physics*).[16] The predictive sciences, however, mostly stayed out of the debate. While relativity and quantum dynamics were serious issues for systems on an atomic scale, they could safely be ignored for systems that were moving at speeds much slower than the speed of light, and that involved a large number of molecules, such as the weather. The mechanistic approach also gained further currency in biology with the discovery of DNA, as will be discussed in Chapter 5.

Somewhat paradoxically, the biological and social sciences, which modelled themselves after the physical sciences, have clung to the deterministic, mechanistic approach long after it was abandoned in atomic physics. The zoologist Richard Dawkins, for example, describes human beings as "lumbering machines" controlled by our genes.[17] The computer scientist Rodney Brooks writes, "We are machines, as are our spouses, our children, and our dogs."[18] But as the physicist David Bohm points out, "At the end of the nineteenth century, physicists widely believed that classical physics gave the general outlines of a complete mechanical explanation of the universe. Since then, relativity and quantum mechanics have overturned such notions altogether. . . . Is it not likely that modern molecular biology will sooner or later undergo a similar fate?"[19]

GOOD VS EVIL

Determinism raises the issue of free will. The ancient Greeks seem to have believed that human destiny was ruled by fate, which could sometimes be divined by consulting with the oracles. With Christianity, it was believed that much of the future was preordained, but that humans had at least some control over their destiny. (Hardliners like the Lutherans, who believed in predestination, were the exception. They felt that since God knows everything, the future had already been decided. Johannes Kepler, a committed Lutheran, was repelled by this notion, and he got in trouble with his church by arguing against it.)

The notion of free will has continued to bounce back and forth like a philosophical tennis ball. Thomas Hobbes, in the mid-sixteenth century, was against it, since he believed that the human mind was ruled by deterministic, cause-and-effect mechanisms. Immanuel Kant, a century later, whacked it back, asserting that at least part of our minds had the capability for free reasoning. Laplace was against it, because of the success of deterministic science, but quantum mechanics reopened the debate. In a system with sensitivity to initial conditions, as the philosopher Karl Popper argued, quantum effects could be magnified to produce a fundamental indeterminacy in our actions. The path of human history could therefore not be predicted or controlled. Popper dedicated *The Poverty of Historicism* to "the countless men and women of all creeds or nations or races who fell victims to the fascist and communist belief in Inexorable Laws of Historical Destiny."[20]

While one interpretation of quantum mechanics implies that matter is fundamentally indeterministic and can be described only in terms of probabilities, another possibility is that it just appears this way to us because we don't understand the underlying dynamics. For example, if an electron is in a particular quantum state, there is a certain probability that it will jump to another state with a different energy level. To us, it may seem that this is a purely free, spontaneous event, but since we cannot claim to understand the full workings of the atom, we don't know if there is a subtle, undetectable force that controls the electron. Does the atom jump, or is it pushed?

If we believe in free will, we might think that when the alarm goes off in the morning, we decide to go to the bathroom, choose to make toast and a cup of tea, and make up our minds what to wear. One could argue, though, that our actions are no less automatic than that alarm going off. Our life is like a movie that is predetermined. It just happens to be very convincingly made, so we go along with it, get into character.

The question of free will has obvious legal implications: if criminals are the product of their environment, then their crimes are not really their fault and the aim of prison should be rehabilitation. People are not intrinsically good or evil. On the other hand, if criminals freely choose their vocation, then justice means retribution rather than therapy.

Science is seen by most of its practitioners as ethically and politically neutral. The scientific process aims to be rational and objective, and therefore eschews subjective value judgments. However, many people choose science over more lucrative careers because they want to contribute to the greater good—by probing the nature of matter, curing disease, or helping the environment. Like priests, they feel they are answering a higher calling. The development of quantum physics, and particularly the discovery that mass can be converted into energy in devices such as atomic bombs, muddied not only the waters of determinism but also the hands of scientists. It's hard to be objective when you have the power to destroy the world. After the bomb was dropped, Einstein said, "If I knew they were going to do this, I would have become a shoemaker."[21] While some physicists did decide to change fields, the invention of the atomic bomb actually led to vast increases in recruitment for high-energy physicists. The most reliable way to get funding for a scientific project is still to show that, like Galileo's telescope, it has a military application.

FEATHERED SERPENT II: RETURN FROM THE DEAD

Nothing does as much to shake one's faith in rational prediction as the arrival of sudden, unforeseen calamities. In Europe, the Great War of 1914–18 destroyed towns, cities, and entire empires. Out of the chaos emerged art movements such as Dadaism and Surrealism, which celebrated irrationality and spontaneity.

The war was followed by an even more deadly event—the flu pandemic of 1918. Known as the Spanish flu, it was actually first recorded in a military camp in Kansas, where in March 1918 it infected over 500 people in two days. The disease spread through the world's population in months, eventually infecting about a fifth and killing tens of millions (estimates vary). Nowhere was spared, with the exception of a few isolated towns and islands, like Marajo in Brazil. Eighteen months later, it disappeared as mysteriously as it had arrived. It was later found to have been caused by an avian virus, normally resident in wild birds, that had mutated to a human-transmissible form. Not as devastating as smallpox was to the Aztecs, but a feathered serpent nonetheless.

Is it possible to predict the course of such pandemics? One method is to assume that the future will resemble the past. In this spirit, scientists in 2005 reconstituted the virus from segments of genetic material recovered from preserved tissue samples of long-dead victims. They hoped to identify predictable mutations in avian flu that lead to virulence and transmissibility in humans. The problem is that the future never does quite copy the past; like financial crises, each epidemic is a little different from the one that came before.

Then again, while nature seems loath to repeat herself, we may do the job for her. The resurrected virus now exists at the U.S. Centers for Disease Control and, in a theoretically weaker version, at a number of other laboratories, and the genetic sequence has been published. This virus is the biological equivalent of an atomic bomb. Of course, it will never escape. Not like SARS, or anthrax, or plague, or ebola—all of which have recently found their way out of secure laboratories. . . .

Our impression of whether a being has free will is often based on the degree to which we find its behaviour predictable. If a system is completely predictable or completely random, then we tend to assume that it is the passive subject of external forces. But if it operates somewhere in between, so that its behaviour has a kind of discernible pattern and order but is still hard to predict, then we believe that it is acting independently. In other words, the autonomy that we assume to be present is a measure of the system's complexity.

COMPLEXITY

Complexity is a branch of science that presents a much stronger challenge to prediction than chaos does. Like most of chaos theory, it developed only in the latter part of the twentieth century with the arrival of fast computers. It began when the Polish mathematician Stanislaw Ulam, following up a 1948 lecture by John von Neumann on the mechanistic modelling of living organisms, proposed the idea of cellular automata. Imagine that you have a two-dimensional grid, divided equally into squares (the cells) that are either black or white. The evolution of the system takes place in discrete steps. In each step, a rule is applied, and that rule determines the colour of each cell. The rules are typically based on the state of adjacent cells. For example, in a system proposed by the mathematician John Conway in 1970, a black cell survives (remains black) in the next step only if it has exactly two or three black neighbours (out of a maximum of eight), and a new active cell is born if it has exactly three black neighbours. Otherwise the cell turns white.

This system is known as Life, for reasons that become evident when you see a simulation (there are many available on the Internet).[22] From a given initial configuration of white and black cells, the screen seems to come alive with strange, shape-shifting forms that are constantly being born, dying, or interacting with one another. Life was invented as a kind of amusing screen saver, but it has since been the subject of much concerted study. Similar automata have been used to describe natural phenomena that also depend on local effects, such as forest fires or earthquakes, as well as more traditional systems, like fluid flow.

The mathematician Stephen Wolfram, who has done much to advance and popularize this area of research, classified the behaviour of such cellular automata into four basic categories:

1. All activity dies out
2. Stable or periodic patterns
3. Apparently random
4. Situated on the border between order and chaos, with no enduring pattern that repeats in a predictable way

The Life system is in the last, and most interesting, category. For these systems, there appears to be a huge discrepancy between the simplicity of the rules and the complexity of the resulting system. In particular, they have a property known as computational irreducibility. Cellular automata can be viewed as performing a complicated calculation: given an initial condition, they will describe the evolution of the system, just as an ODE describes the falling of an idealized stone. However, with Class IV automata, the calculation cannot be speeded up in any way. It's not possible to plug the initial conditions into an equation and crank out the answer. The only way to determine the evolution of the system is to run it and see what happens.

This means that Class IV automata are inherently unpredictable. Again, there is a direct correspondence between a system being interesting—that is, appearing to have a life of its own—and our inability to predict its evolution. Furthermore, according to Wolfram's principle of computational equivalence, once a system has a certain level of complexity—isn't obviously static or periodic or in any other way boring—then it is computationally irreducible.[23]

The rules that define cellular automata can best be described as local or social in nature. Typically, the state of a cell in the next iteration is determined by the current state of neighbouring cells. In Class IV automata, these local rules produce systems that are inherently unpredictable, in the sense that one cannot know the future state of a given cell, but nevertheless have a kind of character, which distinguishes them from automata with other rules. The Life game, for example, has a recognizable zoology of figures, like "gliders" that traverse the screen as if with a mind of their own. These distinctive features are referred to as "emergent properties," which is another way of saying that they just happen. They cannot be predicted or computed from a knowledge of the rules alone—you have to run the system to see them develop. The systems therefore have an intrinsically dualistic nature: they are based on simple, logical rules, but the resulting behaviour is neither simple nor very logical.

Complex systems are like highly decentralized democracies: their decisions are the end result of a large number of local choices. ODEs, on the other hand, are like dictatorships: the physical system passively obeys the laws from central command. In one, the information flows from the bottom up, while in the other, it flows from the top down. Complexity isn't rocket science—it's the opposite of rocket science.

This poses a challenge to numerical prediction because even if a system is deterministic and governed by simple rules, there may exist no mathematical model or set of equations to tell you how it's going to behave—no Apollo's arrow to fly magically into its future. Just as the square root of two cannot be expressed as a ratio of two integers, so a complex system cannot be expressed or predicted by any set of equations. If you start the Life game at a certain initial condition, it will always evolve the same way, but there is no way to tell what a particular cell will be doing after a thousand steps, except by running the game. This is like the weather forecaster telling you that he will have next week's forecast ready by next week. Furthermore, because this unpredictability is not a result of sensitivity to initial condition, it cannot be fixed by improving observations. Often, such systems exhibit only moderate sensitivity to small changes in the initial condition, and even cellular automata that model fluid flow are typically not very chaotic.[24] Rather, the unpredictability is a result of the inherent complexity of the system itself.

This would not be a concern if it was limited to games on computer screens, but cellular automata can emulate a broader range of physical phenomena than ODEs. Since many physical and biological systems are governed by local interactions, there is no a priori reason to assume they can be predicted by the use of equations. As Wolfram writes, "It has normally been assumed that with our powers of mathematics and general thinking the computations we use to make predictions must be almost infinitely more sophisticated than those that occur in most systems in nature. . . . This assumption is not correct. . . . For many systems no systematic prediction can be done."[25] Water is one example of a substance in which simple interactions at a microscopic level lead to complex macroscopic properties. In high school science, a water molecule is typically represented as a stick-and-ball model, with a hydrogen atom connected to two oxygen atoms. The molecule's electrical polarization puts it in constant communication with its neighbours, though, resulting in complex behaviour that could never be deduced from a knowledge of the molecular properties alone. To quote the writer Douglas Coupland, "You can't look at H_2O and predict snow or glaciers or hail or tsunamis, fog, or just about anything else."[26]

While cellular automata can be used to model physical systems, they can usually do so only in a generic sense. Their bottom-up approach relies on a detailed knowledge of local rules, which can never be perfectly known except for idealized, abstract systems. The advantage of equations is that they can approximate the behaviour of a complicated system with simple top-down laws, omitting the details. For this reason, the models used to predict atmospheric, biological, or economic systems are usually based on equations.

Chaos and complexity are often lumped together or confused, but the two are quite different. As mentioned above, complex systems may not be sensitive to initial condition, and conversely, chaotic systems need not be complex. Proof of the latter is the dough-folding shift map, shown in Appendix I.

Of course, if one were really trying to model the making of bread, randomness in the initial condition would be overtaken by other effects: minor variations in the kneading procedure, the texture of the dough, the type of yeast, the fact that the whole thing is put in the oven, and so on. In a word, complications. As the publisher Elbert Hubbard said, life is just one damned thing after another. In a model of the weather, these complications might be referred to as stochastic effects; in a model of the cell, external noise; in economics, random shocks. They arise not from the internal dynamics of the model but from the rest of the world that the model has left out. It is important that models not be confused with this far richer reality.

I will use the term "uncomputable" to describe systems that cannot be accurately modelled using equations without practically recreating the entire system. It may or may not come as a surprise to the reader that physical or biological systems are generally uncomputable, but it certainly comes as a surprise to a fair number of scientists, who have been trained by their institutions to see natural phenomena, from a cloud to a thought, as mathematically tractable. Newton's amazing success at modelling gravity with equations may have inadvertently blinkered subsequent generations. As the cosmologist Hermann Bondi wrote, "His solution of the problem of motion in the solar system was so complete, so total, so precise, so stunning, that it was taken for generations as the model of what any decent theory should be like, not just in physics, but in all fields of human endeavour. It took a long time before one began to understand—and the understanding is not yet universal—that his genius *selected* an area where such perfection of solution was possible."[27]

The predictive sciences have developed around the use of equations such as ODEs not because they are an excellent match for the systems studied, but because they can be solved using mathematics. As we'll see in the next few chapters, we run into severe problems when we use them to model complex, real-world systems. The inherent uncomputability of such systems means that it is extremely hard to do better than the simplest naïve approaches. Models become more and more refined, but as more detail is added, the degree of uncertainty in the parameters explodes. Because the system cannot be reduced to first principles, we must make subjective choices about parameter values and the structure of the equations. The models often get better at fitting what has happened in the past, but they don't get much better at making predictions.

COMPLICATIONS

One winter weekend, when I was an undergraduate student in physics, I was making bread at home. I'd learned the techniques at a summer job in a bakery. Also, my great-grandfather was a baker, so, as they say, it's in my blood. I was kneading the dough and wondering whether the temperature was high enough to allow the bread to rise in a reasonable time. And since my mind was full of physics courses, it occurred to me that in principle, I could build a detailed model that took into account the temperature, the biological state of the growing yeast, the exact state of the dough, and all the other factors, and predict the outcome of the bread experiment using just the laws of physics.

As I pondered this possibility, I realized that an accurate model would have to be as complicated as the system itself. And as I continued to knead, another part of my brain took over—a part that said, Don't be stupid. The dough feels good. (Left brains build detailed mathematical models of systems that involve the interplay of atmospheric and biological components; right brains just sniff the air or give a squeeze and come to their own conclusions.)

Rolling the moist dough over the floured kitchen surface, I realized that while building models was an impressive and fascinating task, it could also have negative consequences, because it interferes with a person's direct experience of an event. Instead of enjoying the smell of the bread and the sensation of forming the dough, I would become obsessed with controlling the timing and the temperature and the measurements of the ingredients. It seemed that my physics education was acting like a kind of indoctrination, forcing me to see the world in this very controlling way. Partly as a result of this dramatic flash of insight, I dropped the physics portion of the program and switched to mathematics, which leaves applications up to the user—and also allowed me to take film studies as an option.

When he wasn't laying down the foundations of Western European thought, Aristotle also developed the theory of how to write screenplays. His *Poetica*, standard reading in film courses, describes the basic structure of a three-act Hollywood movie. Every story has a beginning, a middle, and an end, but not necessarily in that order (Jean-Luc Godard). The first act sets up the characters and the situation—the initial conditions. In the second act, the arc of the story is diverted by friends, enemies, accidents—the complications. There is a fluid buildup of tension and conflict, as forces grow and align against one another. In the third act, the drama is resolved and peace is restored, but—if the story works—in a not altogether predictable way.

Life without a little conflict, complexity, or complication is dull. A common criticism of Hollywood movies is that they are too formulaic; we've seen it all before, and we know exactly what is going to happen. Part of the enduring appeal of Shakespeare's plays is that they elude tidy formulation. In the original story on which *Hamlet* is based, the action begins when Hamlet is only an adolescent and it becomes publicly known that his uncle murdered his father. Hamlet plays the fool for several years, pretending to be mad so that his uncle does not see him as a threat, while secretly waiting until the time is right for him to revenge his father and claim the throne. In Shakespeare's version, the action is compressed into days, and now Hamlet's feigned madness only serves to draw attention to him. The external, transparent motivation is replaced by inner compulsions; the characters seem like real people, not just cogs in a machine.[28]

This points to the fundamental danger of deterministic, objective science: like a corny, over-formulaic film, it imagines and presents the world as a predictable object. It has no sense of the mystery, magic, or surprise of life. A key difference between a living thing and an object is predictability: kick a stone, and you know what will happen; swat at a bee, and things get more complicated. By objectifying nature, we kill it and demystify it, as surely as Apollo's arrows ran through Python. Galileo's argument against the sterility and "immutability" of the Aristotelian world-view can be turned around. As the psychiatrist R. D. Laing wrote, "Galileo's program offers us a dead world: Out go sight, sound, taste, touch, and smell, and along with them have since gone esthetic and ethical sensibility, values, quality, soul, consciousness, spirit."[29] A consequence is that scientists have come to believe that our weather, our economy, and even our own bodies should be as predictable as a falling stone.

Indeed, the attentive (or merely skeptical) reader may have noticed that in this book about prediction, the only thing we have so far shown how to predict is the motion of objects like planets, stones, and apples (even bread dough is beyond us). For simple systems, it is possible to write down equations that accurately describe their motion. So what techniques are available when we want to predict a real system we care about, like the weather? Pretty much the same thing. We write out sets of equations, code them in a computer, measure the initial conditions, and launch. Does it work? Not as well as with falling objects. One could say that Newton's apple is the low-hanging fruit of predictability. Real-world systems, even ones that most people wouldn't consider to be alive (such as the weather), appear to be uncomputable. As we'll see in the next three chapters:

- **Predictive models are based on sets of equations.** These attempt to simulate atmospheric, biological, or economic systems using a top-down approach that omits details, is amenable to mathematical analysis, and tells an understandable story.
- **However, the underlying systems cannot be reduced to equations.** They are based on local rules, and their global, "emergent" properties cannot be computed.
- **Models of these systems tend to be sensitive to changes in parameters.** The models can be adjusted to fit past data, but this does not mean they can predict the future.
- **More data and bigger computers do not necessarily help.** Adding more detail often gives diminishing returns, because the number of unknown parameters explodes.
- **Statistical methods can sometimes be of use.** However, these are often based on vague correlations, rely on the future's resembling the past, and do not provide a cause-and-effect explanation.
- **Even simple models can sometimes be used to make predictions.** These usually take the form of subjective remarks or warnings, rather than precise forecasts.

The history of prediction presented so far is a low-resolution model that picks up only the coarse features. It addresses our impulse to explain the world in terms of linear cause-and-effect relationships. Apollo (if the stories were true) begat Pythagoras begat Plato begat Aristotle begat Kepler begat Galileo begat Newton begat Einstein. Of course, it wasn't really like that. Context matters. None of those people, demi-gods, or gods existed in a vacuum. They all had colleagues, societies, families, wives (but no husbands—the oracle's sex change, as Ralph Abraham called it, was not reversed).[30] However, telling the story in these simple terms helps us to organize and understand what happened, to see where we are coming from. The same linear, cause-and-effect method doesn't always work when we go in the other direction and try to predict the future.

PRESENT

4 ▸ Red Sky at Night
Predicting the Weather

Big whirls have little whirls that feed on their velocity, and little
whirls have lesser whirls and so on to viscosity—in the molecular
sense.
   —Lewis Fry Richardson, British physicist and psychologist

There are known knowns; there are things we know we know.
We also know there are known unknowns; that is to say, we
know there are some things we do not know. But there are also
unknown unknowns—the ones we don't know we don't know.
   —Donald Rumsfeld, secretary of defense of the United States

Taking the Temperature

If the world has a favourite opening topic for conversation—an
icebreaker—it is probably the weather. If you are ever lost for
anything more meaningful to say, you can always try "Chilly out,
isn't it?" (or the equivalent in the local dialect). It usually isn't very
controversial, unless it's noon in the Sahara, and it might lead to
something more interesting.

   One of the first to write down his thoughts on the atmosphere
and the oceans was, again, Aristotle. Like a good scientist, he focused

on hammering the theories of his competitors and promoting his own. "The belief held by Democritus that the sea is decreasing in volume and that it will in the end disappear is like something out of Aesop's fables," he thundered in *Meteorologica* (meteorology was defined more broadly then, and encompassed such phenomena as comets, earthquakes, and oceans). "Plato's description of rivers and seas in the *Phaedo* is impossible," he declared. "It is equally absurd for anyone to think, like Empedocles, that he has made an intelligible statement when he says that the sea is the sweat of the Earth. Such a statement is perhaps satisfactory in poetry, for metaphor is a poetic device, but it does not advance our knowledge of nature." His own view was that thunder, lightning, and hurricanes were all caused by "windy exhalations," perhaps from quarrelling philosophers.

Aristotle's student Theophrastus of Eresos compiled the first attempt to predict weather using empirical observations. It included such chestnuts as "Red sky at night, sailor's delight / Red sky at morning, sailor take warning," and created a basis for weather forecasting that, while short on theoretical backing, persisted for two millennia.

Like scientific prediction in general, weather forecasting didn't take a major step forward until the Renaissance, when Galileo built the first thermometer and his student Torricelli followed up with the barometer. Galileo's early version of a thermometer was a glass bulb attached to a tube the size of a straw. In lecture demonstrations, he would warm the bulb and insert the end of the straw into a basin filled with water. As the bulb cooled, the air inside would contract, drawing liquid up into the straw. The height it reached reflected the room's temperature, so Galileo could make it go up or down by moving the device from a warm place to a cold one.

A variety of improved versions soon appeared. One, by Robert Hooke, was filled with "best rectified spirit of wine highly ting'd with the lovely colour of cochineal," perhaps so that you could drink the contents if it got too cold out.[1] In 1714, Gabriel Fahrenheit constructed sealed mercury thermometers with reliable scales, which turned the device into a serious scientific instrument.

Temperature, we now know, is a measurement of the kinetic energy of air molecules—that is, how fast they are moving on average—as opposed to wind, which measures the net overall motion of a parcel of air. The barometer measures the atmospheric air pressure, which is essentially the air's weight, or number of molecules per unit volume. Pressure and temperature therefore represent statistical averages over a large number of particles. It is meaningful to take such averages because one air molecule is assumed to be physically the same as any other.

Aristotle tried to weigh air by first weighing a leather bag when it was pressed flat, then weighing it again when it was full of air. He found there was no difference, so concluded that air was without weight. Galileo, always the skeptic, tried to devise better experiments, but it was his student Torricelli who managed to accurately weigh air with the barometer. This was a metre-high tube partially filled with mercury, its open end inverted into a reservoir so that the weight of the air pressing down on the reservoir was balanced by the mercury in the tube. Fluctuations in pressure caused the mercury level to go up or down.[2] Blaise Pascal proved that pressure decreases with height by convincing his brother-in-law to lug a barometer up a mountain and compare the pressure at the top with that at the base. (The air at the bottom has all the weight of the air above pressing down on it, which increases the pressure.)

The fact that air has weight will seem reasonable to anyone who has seen tornadoes tossing around telephone poles like they were matchsticks. But of course it's not just the weight, or the heat, that matters: it's the humidity. Molecules of water can be mixed in with the air. An early hygrometer, for measuring humidity, was supplied by the ever-inventive Hooke, who employed the "beard of a wild goat." This protruded from a hole in the top of a well-ventilated box, and its end was attached to a pointer. As the hair curled or uncurled depending on the moisture in the air, the pointer indicated the humidity on a dial. Similar hair hygrometers are still in use today, though another method is to compare the temperature of the humid air with that of dry air.[3]

By the mid-1600s, temperature, pressure, and humidity were being measured on a regular basis. In 1724, the Royal Society managed to organize a truly international data-collection arrangement, complete with standardized forms and instruments. Synchronized observations had to await the invention of the telegraph in 1844—until then, the weather tended to outrun the messenger—and led to weather predictions based on a synopsis of the available data. The synoptic method exploited the fact that the weather has identifiable features, such as masses of warm/cold or dry/humid air, or storms. If, say, you learn over the telegram that frigid polar air was last seen heading your way, then it is reasonable to suppose that it will soon get cold.

FORECAST

The first meteorological office was set up in Britain in 1854. The Met. Office, as it became known, was headed by Admiral Robert FitzRoy—he who had captained the *Beagle* and taken Darwin around the world. The ex-navy man saw that weather forecasting had the potential to save lives by warning mariners of storms, like the one that destroyed nearly thirty French and British vessels in the Crimean a month before his appointment. A network of forty weather stations was set up around the United Kingdom, and weather reports were published in London newspapers. In France, the chemist and accountant Antoine Lavoisier funded a chain of observation stations before being sent to the guillotine for his unpopular taxation activities.

FitzRoy's efforts were also not well received, by the public or the scientific establishment. At the time, weather prediction was something practised by astrologers, and it was not seen as a fit subject for scientific pursuit. The popular press enjoyed comparing the Met. Office's inaccurate predictions to those from astrological sources, such as Zadkiel's Almanac. The mainstream scientists saw all this as a threat to their reputation.

FitzRoy tried to blunt the comparison to astrology by avoiding loaded words like "prediction"; instead, he invented a new word of his own: forecast. "Prophecies or predictions they are not; the term forecast is strictly applicable to such an opinion as is the result of a scientific combination and calculation."[4] In 1863, he published *The Weather Book,* which tried to make the weather comprehensible to people of average education. But his attempt to popularize the subject further annoyed elitist scientific institutions like the Royal Society. It didn't help that he spoke about the weather in intuitive rather than mathematical terms, and claimed that the telegraphic network had provided him with a "means of feeling . . . successive states of the atmosphere over the greater extent of our islands,"[5] which sometimes sounded more mystical than objective.

On April 30, 1865, at the age of fifty-nine, Robert FitzRoy took his own life by slitting his throat with his razor. He might have had an inherited tendency to depression. His uncle Lord Castlereagh had similarly killed himself, and Darwin, who dined with him on the *Beagle* every day for five years, had noted his occasionally stormy temper. He might also have been affected by his association with Darwin's theory of evolution, which, as a creationist, he considered blasphemous. However, it appears that the primary cause of his depression was being caught between the so-called astro-meteorologists on the one side and the scientific establishment on the other. After his death, a committee formed by the Royal Society and the Board of Trade, and chaired by Darwin's cousin Sir Francis Galton, released a report that tore apart every aspect of FitzRoy's work. It claimed that his forecasts were not deduced "by means of accurate induction from known facts," and were "wanting in all elements necessary to inspire confidence."[6] As a result, storm warnings were suspended.

FitzRoy, however, was not without supporters. Fishermen, maritime insurers, and the navy had actually found the storm warnings useful, and they were reinstated in 1867. Public interest in weather forecasts was also aided by the publication, in the *London Times,* of weather maps, which like their modern versions showed winds, precipitation, temperature, and air pressure.

OBSERVATION AND THEORY

Galton had become interested in maps and meteorology while exploring southwestern Africa for the Royal Geographical Society, and he played a large part in the development of weather maps. He was nothing if not an enthusiastic measurer. On his many journeys to Africa and elsewhere, he measured the longitude and latitude of towns and the altitude of mountains (by noting the barometric pressure at the summit). On one occasion, he measured an African woman whose voluptuous physique he found particularly striking. He recorded "a series of observations upon her figure in every direction, up and down, crossways, diagonally, and so forth, and I registered them carefully upon an outline for fear of any mistake; this being done I boldly pulled out my measuring tape, and measured the distance from where I was to the place where she stood, and having thus obtained both base and angles, I worked out the results by trigonometry and logarithms."[7] As we'll see in the next chapter, his interest in measuring human bodies continued when he later shifted from meteorology to his theory of inheritance.

The Norwegian scientist Vilhelm Bjerknes also believed that meteorology needed to be shed of its astrological associations and turned into a hard science like physics. The key would be to show that it could provide accurate predictions. Echoing Laplace's all-seeing demon, he wrote in a 1904 paper that this would require two things: "1. A sufficiently accurate knowledge of the state of the atmosphere at the initial time. 2. A sufficiently accurate knowledge of the laws according to which one state of the atmosphere develops from another."[8] In other words, forecasting requires the initial condition (today's weather) and a model of the atmospheric dynamics. Observation and theory; a Tycho and a Kepler.

As it happened, the quality of both these factors was improving because of the impending war in Europe. Regular flights by airships and airplanes meant improved observational data. The new military technologies of aviation, long-range artillery, and gas warfare were all affected by the weather. Meteorologists suddenly found themselves in demand by military planners. Indeed, the term "front" for the boundary between different air masses was taken from First World War military terminology.

Bjerknes devised a set of seven differential "primitive equations," based on the Navier-Stokes fluid-flow equations from the 1840s, to model the atmosphere. They were essentially an application of Newton's laws of motion to fluid flow, and they described how the atmospheric variables of temperature, wind, air pressure, and humidity change with time. Some of the relations can be understood intuitively. For example, differences in pressure between regions create a force that drives winds from high pressure to low. Converging winds increase the pressure, while diverging winds decrease it. Temperature also varies with pressure, since air gets warmer as it is compressed. The equations could only coarsely approximate the effects of turbulence, or the behaviour of water vapour, but they captured the overall atmospheric flow.

Solving the equations numerically was difficult, so Bjerknes had to rely on graphical techniques, based on weather maps, to find approximate solutions. However, this was so slow that it was hard to make useful forecasts before the weather arrived. As the Englishman Lewis Fry Richardson realized, true numerical weather prediction would require computers—though he used the word differently than we do today.

THE STORMS OF WAR

Richardson was a Quaker and a pacifist. When the First World War started, he had a comfortable position working for the Met. Office at Eskdalemuir Observatory in Scotland. In 1916, at the age of thirty-five, he quit his job to serve as an ambulance driver in the war. While helping ferry the wounded away from the fronts, he developed the ideas behind his book *Weather Prediction by Numerical Process.* His aim was to show how the weather could be predicted using the method of finite differences.

Richardson had experience with the finite difference technique from a previous position with the National Peat Industries, where he had modelled the flow of water in peat. The method works by dividing both space and time into a Cartesian grid. At the initial time $t = 0$, the values of atmospheric variables such as temperature, wind, and pressure are specified at each grid point. The finer the grid, the greater number of variables. At time Δt in the future, the differential equations are used to anticipate how the variables will change, based on the physical laws. This step is repeated each Δt time units. The procedure is similar to the numerical technique used for solving Newton's laws of motion, described in Chapter 2, except that instead of a single falling stone, the system was a swirling mass of air and vapour.

For a square area measuring about 1,000 kilometres a side, roughly centred on Germany, Richardson chose a grid of 5 divisions in each horizontal direction and 5 vertically, for a total of 125 cells. Each was about 200 kilometres square and 2 to 5 kilometres high (since pressure decreases with height, the upper layers were made taller so they contained a similar mass of air). For his initial condition, he used observations provided by Bjerknes for May 20, 1910, when observatories throughout Western Europe had released a large number of weather balloons.

Because the balloon launches did not coincide with the grid, Richardson's first step was to estimate, or interpolate, the values for each cell based on the nearest measurements. He then attempted to calculate changes in the atmospheric variables six hours into the future, using a time step Δt of three hours. Since there were hundreds of variables, and even more equations, the computation was a massive task. It was amazing enough that Richardson completed it, even more so that he did it while in the middle of a war. Scribbling calculations into his notebooks, his office a "heap of hay in a cold rest billet," he almost lost everything during the Battle of Champagne, in April 1917, when the manuscript was sent to the rear and apparently lost, before being found months later under a heap of coal. In the end, he succeeded in producing the world's first numerical forecast (or hindcast, since the date was in the past). Unfortunately, his predicted changes in air pressure were out by an ear-blowing factor of about a hundred. He blamed poor observational data, though instabilities in his numerical method also played a role.

The calculations for a six-hour prediction took him about the same number of weeks to carry out. "Perhaps some day in the dim future it will be possible to advance the calculations faster than the weather advances," he concluded. "But that is a dream."[9] He imagined a large hall, a kind of "forecast factory," full of "computers," by which he meant people armed with slide rules, each of whom would be responsible for calculating the changes in one small cell of the atmosphere. The walls would be painted with a map of the globe, and at a pulpit in the centre would be a single person, "like a conductor of an orchestra," who kept everybody in time. Richardson estimated that his scheme would require 64,000 "computers" just to keep pace with the weather. Perhaps as a result, other scientists didn't try to pursue numerical weather forecasting.

Another approach, pioneered by the climatologist Sir Gilbert Walker in his attempts to predict the Indian monsoon rains, was to look for statistical patterns in past data.[10] That didn't catch on either—in part because of the lack of high-quality, global observational data; in part because such statistical methods do not provide a testable, cause-and-effect explanation for phenomena; but mostly because weather patterns are rather unreliable.[11] However, the techniques Walker developed to find patterns and correlations in graphs of variables such as rainfall versus time, known as time series analysis, helped to detect weather patterns associated with El Niño and are still used in many areas of prediction, including finance.

LONG-DISTANCE CONNECTIONS

El Niño is now usually associated with extreme weather, but the name was first used to describe a benign event off the coast of Peru—a warming of the ocean waters. This, it was said, would lead to an *año de abundancia,* when "the sea is full of wonders, the land even more so. First of all the desert becomes a garden. . . . The soil is soaked by the heavy downpour, and within a few weeks the whole country is covered by abundant pasture. The natural increase of flocks practically doubles and cotton can be grown in places where in other years vegetation seems impossible." (Murphy 1926). The inhabitants of the region called the warm current El Niño, after the Christ child, since it usually set in shortly after Christmas. (The opposite state is now called La Niña, "the little girl.")

It sounds like something out of a Gabriel Garcia Márquez novel—a magical flowering of the desert, a purely local event. It was only relatively recently that the phenomenon was linked with other, often disruptive, happenings around the world. In fact, El Niño is the ocean part of a global weather pattern that also includes the Southern Oscillation, a see-saw fluctuation in atmospheric pressure between South America and India/Australia that was first reported in 1923 by Sir Gilbert Walker when he was director general of the observatory in India. The coupled ocean/atmosphere phenomenon, known as ENSO, causes everything from droughts in Amazon rainforests to monsoons in India.

ENSO provides an example of how seemingly unrelated events can be part of a larger pattern. Such teleconnections, as meteorologists have termed them, blur the line between cause and effect, and mean that the ocean and the atmosphere have to be treated as a single system.

In 1920, the Met. Office became part of the Air Ministry, so Richardson would have become a military employee. He wasn't interested in modelling the flow of poison gas or being involved with the military in any way, so he resigned. He took a teaching position and signed up as an external student at University College London to study the new science of psychology. His research from then on focused not on meteorology but on something he considered far more important: the dynamics of war. Richardson believed that science was subordinate to morals, and he was motivated not just by intellectual satisfaction but by his pacifism and what he had seen during his ambulance-driving days. He modelled the buildup of conflicts the same way he had once modelled the buildup of a storm, by using differential equations.[12] These models were almost a kind of fable. They showed how, like a hurricane developing from a small disturbance, an arms race between two countries can rapidly spiral out of control, as had happened in the First World War and would happen again in 1939.

THE GCM

Richardson's dream of numerical weather prediction was eventually realized in the 1950s, albeit with newly invented silicon computers, not human ones. He would have been less pleased by the fact that most of the new technology was developed as a result of military efforts during and after the Second World War. The meteorological observation network had been expanded to include thousands of balloons radioing back information about temperature, humidity, wind speed, and pressure. High-speed computers were developed to crack enemy codes and model the explosions of atomic bombs. The brilliant Princeton mathematician John von Neumann, who cut his teeth on quantum mechanics and worked during the war on the non-linear dynamics of thermonuclear explosions, realized that the fluid dynamics of atom bombs could be applied to model the atmospheric flow.

Most meteorologists were skeptical. The punch-card computers of the time had nowhere near the speed of a modern desktop machine, and it seemed unlikely that they could perform the complex calculations required. Henry G. Houghton, the president of the American Meteorological Society, said in 1946, "There appears to be no immediate prospect of an objective method of forecasting based entirely on sound physical principles."[13] Forecasters, it was believed, would always have to rely on what Carl-Gustaf Rossby, Houghton's predecessor, had called "the horrible subjectivity."[14]

But the skeptics didn't count on two things: huge increases in computing speed, and the simplification of the equations by mathematicians such as von Neumann's new hire, Jule Charney. Charney's quasi-geostrophic approximation replaced Bjerknes's seven equations with a single one. This still had to be evaluated at each three-dimensional grid point, but it made the atmospheric problem computationally tractable, especially since it filtered out the high-frequency oscillations that tended to make models unstable.

In 1950, numerical weather prediction was demonstrated on a computer at the U.S. Army's Ballistic Research Laboratory in Maryland. The regional model divided North America into 270 two-dimensional cells. Running on a punch-card machine known as ENIAC—for Electronic Numerical Integrator and Computer— the twenty-four-hour forecast took about twenty-four hours to complete, but the result at least resembled the actual weather. The first big success came in November 1955, when a weather model beat human forecasters in accurately predicting a storm in Washington, D.C.

Inspired by this, Joseph Smagorinsky and Syukuro Manabe at the U.S. Weather Bureau set about building a three-dimensional model of the global atmosphere. Based on Bjerkness's primitive equations, it included such details as how the atmosphere exchanges water and heat with the planet's surface, and how the hydrological cycle (in which rain falls to the ground and re-evaporates) works.[15] The GCM, which stood for general circulation model (and later for global climate model, or global coupled model—but not for Greek Circle Model), was born.

Other teams soon set to work on their own GCMs. Finally, it seemed that Richardson's dream, and even the deterministic vision of Laplace, was within reach. If we could accurately measure the current state of the atmosphere—Bjerkness's initial condition—and apply the physical laws—the GCM—then we could predict the future weather as surely as we could the trajectory of the moon around the earth. And we could also control it. Hurricanes could be tamed or steered away, clouds seeded to produce rain, floods avoided.

The military potential of forecasting was also obvious. The outcome of battles has often depended on the weather. In 1588, the Spanish Armada lost more ships to terrible September storms than it did to the British navy. The Spanish should have heeded the astrologists, who had been predicting bad things for that year: "Total catastrophe may not occur, but the storms will cause havoc by land and sea and the whole world will suffer astonishing upheavals, followed by widespread sorrow."[16] The Allied invasion on D-Day was delayed for a day when forecasters accurately predicted a storm that would have disrupted the landing. Like Apollo's arrow, forecasts have always been used as an instrument of war.

Von Neumann even believed that the weather itself could be manipulated and turned into a weapon. To this end, the RAND corporation, a defence think-tank in California, investigated methods to alter local climates.[17] The weathermen would be rainmakers. They might not have the bomb, but at least they could inflict a drought on their enemies or depress the hell out of them with extended periods of cold.

In a United States intent on winning the real Cold War, resources continued to pour into meteorology and computer science. With the establishment of academic programs focused on the study of atmospheric dynamics, such as the meteorology department at the Massachusetts Institute of Technology (MIT), meteorology realized its aim of being viewed as a serious, objective, "hard" science. While GCMs were going up in the world, though, they still couldn't predict the weather over longer time scales. As a panel of the U.S. National Academy of Sciences dryly reported in 1965, the best models could simulate the atmosphere with features "that have some resemblance to observation." They concluded that more computer power was necessary.[18]

However, there was another fly in the ointment—or rather, a butterfly. Just as Poincaré had done in the previous century, when he realized that the three-body problem was sensitive to initial conditions, the MIT meteorologist Ed Lorenz discovered chaos. The difference was that unlike Poincaré, Lorenz could visualize the solutions to his equations on his computer screen—and they looked like a butterfly.

STRANGE ATTRACTION

In the early 1960s, Lorenz had been working on a highly simplified model of atmospheric flow, involving only a handful of equations. On one occasion, he interrupted a simulation before it had finished and wanted to restart it. The computer program was working to six-digit accuracy but outputting only three digits, so he restarted from the three-digit numbers and went for coffee while the machine, which was built from vacuum tubes, cranked away. When he returned, he found that the new "weather" output, or trajectory, was completely different from the previous version. After checking for mistakes, he realized that the difference was the result of the round-off in the initial condition. This sensitivity to initial condition was highly disconcerting. It was as if Newton, on seeing the apple fall from the tree, had repeated the experiment by dropping the fruit from almost the same position, and found that it shot off in a different direction.

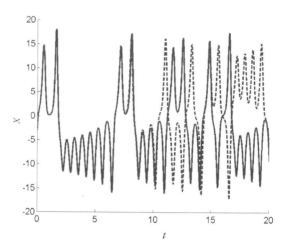

FIGURE 4.2. Plot of the variable *x* versus time *t* for the Lorenz system. The solid line shows a trajectory (i.e., a simulation of the system) initiated at a point on the attractor; the dashed line shows a trajectory initiated at the same point but rounded to three digits. The small error in initial condition results in a trajectory that clearly separates from the first after about ten time units.

He later produced similar behaviour using an even simpler model, which simulated convective flow. When a container full of air is heated from underneath, the warm air at the bottom rises to the top, where it cools, causing it to sink down, and so on in a loop.

Similar convective flow occurs in the atmosphere, when air warms in the equator, for example. Lorenz's model, though, was highly abstract and was not intended to accurately resemble the real physical dynamics (see Appendix II for details). In Figure 4.2, the solid line is a plot of the variable x as a function of time. It oscillates for a while at a high level (0 to 2 on the horizontal time scale), then switches to a low level (2 to 6), and continues back and forth at apparently random intervals. If a second trajectory (the dashed line) is initiated at the same initial condition but rounded off to three decimal places, then the small error grows until the two trajectories become clearly distinct after about ten time units, just as Lorenz found.

The order in this chaos becomes more clear when two variables are plotted against each other, as in figure 4.3. The trajectory then forms a shape with two lobes that resemble the wings of a butterfly. In dynamical systems theory, this is referred to as an attractor, since no matter what initial condition is used, the trajectory will end up on it. The system is therefore bound and limited by the attractor. The butterfly is not free but pinned.

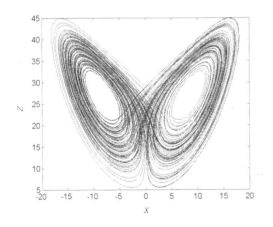

FIGURE 4.3. A plot of the variable z versus x reveals the butterfly-shape attractor of the Lorenz system.

There are three basic types of attractor. In a point attractor, trajectories are drawn to a single fixed point. An example is a pendulum. There, the attractor is the state where the pendulum points straight down. If it is perturbed slightly, it will swing back and forth, with the amplitude of each swing decreasing because of air resistance and friction until it comes to a halt. In a periodic attractor, trajectories are drawn into a repeating cycle, like the lightly forced pendulum in an old-fashioned clock. The third class, to which the Lorenz system belongs, is the so-called strange attractor, which is characteristic of chaotic systems and has a more complex appearance. (The word "strange" does not imply that these attractors are unusual, only that they were discovered after the other two types.)

In keeping with our baking analogies, we can look at the Lorenz system as a kind of Mixmaster with twin beaters, one for each lobe. A particle in the mix circulates around one of the two lobes, and occasionally switches to the other. The mix is being stretched by the beaters, so two particles that start off close to each other are quickly pulled apart (as one would hope, since it is a mixer). The spread of the mix is limited by the bowl. The effect is very much like the shift map of Chapter 3: the stretching of the dough means that the distance between particles initially increases by a factor of two at each step, but it's limited because the dough is constantly folded back onto itself.[19] The main difference is that the Lorenz system is continuous in time, while the shift map is specified only at discrete times.

Like the shift map, the Lorenz system also shows exponential growth of errors. This can be seen by plotting the distance between one trajectory and a second from a perturbed initial condition. We first need a measure of distance in three dimensions. In a two-dimensional Cartesian grid, the distance between a point with the co-ordinates x, y and a second point x_p, y_p can be obtained by using Pythagoras's theorem (as shown in figure 4.4).

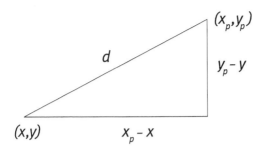

FIGURE 4.4. Calculating distance in a Cartesian grid. The horizontal distance is $x_p - x$, while the vertical distance is $y_p - y$. From Pythagoras's theorem, the total distance d is therefore given by $d = \sqrt{(x_p - x)^2 + (y_p - y)^2}$.

Similarly, the distance, or error, between a point on one three-dimensional trajectory to a point on the perturbed trajectory can be found by squaring the error in each co-ordinate, summing the squares, and taking the square root of the result. The evolution of the error will depend on the initial conditions of the trajectories; however, we can get the expected error growth by performing a large number of experiments and taking the average. Actually, it is more common in meteorology to use the root mean square (RMS) error, which is based on the square root of a sum of squares, instead of the average, though the two have similar properties.[20] As seen in figure 4.5 (on page 142), the RMS error increases in a quasi-exponential fashion over the first time unit. As in the shift map (see also figure A.1 in Appendix I on page 352), the error growth eventually saturates; all trajectories must stay on the attractor, so the distance between them is limited.

The Lorenz system was not intended to be an accurate representation of convection—it was a truncation to three equations of a larger system, which itself was only an approximation and did not fully account for viscosity, which would tend to damp out small perturbations.[21] However, it did show in principle that atmospheric flow-type systems could behave chaotically. Since the initial condition—the exact current state of the atmosphere—could never be perfectly known, it followed that we could never perfectly predict the future weather, even if the model was perfect. Errors grew exponentially, so it was only a matter of time before they became

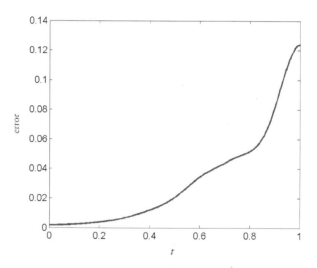

huge.

FIGURE 4.5. Plot of RMS error as a function of time for the Lorenz system, for an initial perturbation of magnitude 0.001 in each variable.

BLAME THE BUTTERFLY

One would think that such a Malthusian statement would be poorly received, and indeed the initial reaction was muted. However, Jule Charney, who worked down the hall at MIT, realized the importance of sensitivity to initial condition: perhaps it could account for the poor performance of weather models. Charney therefore decided to determine the rate of error growth in real GCMs, and he asked meteorologists from ten countries to repeat Lorenz's experiment by initiating trajectories from slightly different points and seeing how quickly they diverged.

The fact that errors grow in an exponential fashion does not necessarily mean they grow quickly. The money in my checking account grows exponentially, but since the interest rate is less than a percent, I would have to wait about a hundred years for it to double, and rather longer to become a millionaire. A useful measure is the doubling time. If weather errors double in magnitude once every day, then after one week, they would have increased by a factor of $2^7 = 128$. On the other hand, if the doubling time was one week, they would increase by only a factor of two in that time. In fact, if errors grow exponentially, rather than linearly, over the first week, they are actually smaller for times less than one week.

It turned out that the average doubling time from Charney's experiments was about five days.[22] This wasn't particularly fast, but it did effectively put a lid of about seventeen days on accurate forecasts, since any errors would by then blow up by about a factor of ten. Also, the growth of initial errors depended to an extent on the type of perturbation that was made to the initial condition, and it was hard to know what would be realistic. As we'll see, later reports gave much faster doubling times, down to about one day. As models increased in resolution, dividing the atmosphere into a finer and finer grid, employing as many as 10 million variables, it seemed their behaviour grew more chaotic.

But if anything in science was growing exponentially, it was interest in chaos theory. Chaos (which is a fascinating area of mathematics) was running amuck not just in meteorology, but in many other branches of applied science, such as biology and economics.

Variations in populations of species or the beat of a heart or the stock market were being modelled as chaotic systems. Interest was fuelled by advances in computing, which meant that for the first time, scientists could rapidly solve differential equations and visualize the solutions on desktop machines. Many non-linear systems, such as the Mandelbrot set, turned out to have beautiful fractal properties with amazingly rich detail, no matter how closely one zoomed in. The phenomenon of chaos got even more attention in 1972, when Lorenz gave a talk at the American Association for the Advancement of Science and introduced the catchy term "butterfly effect" to describe the sensitivity to initial condition. By the early 1990s, the eggs this butterfly had laid in scientists' minds were fully hatched: "It is the errors that arise due to instabilities in the atmosphere (even in case of small initial errors) that dominate forecast errors," concluded one paper.[23]

While it might have been depressing to some that chaos had put perfect weather forecasting out of reach, the story it told—the subtext of the equations—was not altogether negative. It helped answer the question of why forecasts always went wrong, despite the massive amounts of time, resources, and brainpower that were expended on them.[24] The problem was not with the deterministic approach, the physical laws the GCMs were based on, or the quality of the science, but instead was a natural consequence of the equations themselves. Furthermore, while the butterfly effect meant that exact predictions of the future weather were impossible, this had no effect whatsoever on calculations of the long-term climate. Just as any trajectory in the Lorenz system will eventually settle onto the attractor, so calculations of a climate model's "attractor" are not dependent on the initial condition.

In any case, the existence of chaos did not imply that forecasts should not be made—only that there should be more of them. Weather forecasting centres such as the European Centre for Medium-range Weather Forecasts (ECMWF) in Reading, England, and the National Center for Atmospheric Research in Boulder, Colorado, began developing elaborate ensemble forecasting techniques to deal with the effects of chaos. If nearly all errors were due to the initial condition, then a number of forecasts from perturbed initial conditions could be used to form a probabilistic forecast. The ensemble teams were even comfortable enough with the models to adopt a perfect model assumption: "We will assume that our numerical model is essentially perfect," wrote one.[25] A strange thing happened, though. Instead of diverging rapidly, as expected, the perturbed forecasts tended to cluster closely together. It was decided that these were the wrong kind of perturbations, and sophisticated algorithms were developed to find specialized perturbations that diverged more rapidly.

The butterfly therefore remained pinned down as the source of practically all forecast error. But is the weather really so delicate and finely poised a system that an insect can stir up a hurricane—or knock it off track—with a beat of its tiny wings?

Our Daily Forecast

The fundamental causes of atmospheric motion are relatively simple and relate not so much to butterflies but to that other dynamical system that Pythagoras called the cosmos. If the earth were stationary relative to the sun, so that one side always faced the sun and the other side faced away (as the moon does with the earth), then the weather would be extremely dull. One half of the planet would have perpetual daylight, while the other had perpetual night. One side would be warm, the other cold. Viewed as a dynamical system, the weather would be almost like a pendulum at rest.

Instead, the earth is constantly spinning like a top, which

means that each side alternates between day and night, warming and cooling. Also, the planet rotates around the sun, with a period of one year. Because the earth's axis is at an angle to the sun (as in figure 1.3), first one hemisphere, then the other gets more exposure to light. This results in the seasons. The atmosphere reacts to this differential heating by attempting to equalize the temperature. Warm air at the equator lifts up, and cold air from the poles moves in to replace it. The atmosphere is constantly being churned around and can never reach equilibrium.

The flow of air is affected by the fact that the planet is round and is spinning. Land at the equator is travelling fast, but land closer to the poles moves more slowly (because the radius of rotation is smaller). As soon as air begins moving—say, from west to east—it interacts with the spinning motion to curl from right to left in the northern hemisphere and left to right in the southern hemisphere. This is known as the Coriolis effect, and it leads to the swirling patterns of winds seen on weather reports. Also, air is subject to the non-linear effects of friction and turbulence, especially at lower latitudes, where it interacts with the land and the oceans. And then there are the effects of moisture to consider: water picked up from the oceans forms clouds, which affect the local heating, which affects the wind, and so on.

The first step in weather prediction, as Bjerknes wrote, is to specify the initial condition: in other words, today's weather, as measured by the basic variables of temperature, pressure, wind speed and direction, and humidity. Because it isn't possible to know these variables at every point, the atmosphere is divided into a giant three-dimensional grid, the spherical counterpart to Cartesian coordinates, which surrounds the world like a cage. The resolution—the coarseness of the division—is determined only by the available computer power. In a modern GCM, the resolution might be about forty kilometres horizontally and one kilometre vertically, though this depends on the exact model. (Local models that simulate the weather in particular geographical regions often use a finer grid but rely on global models for inputs.)

Because the atmospheric variables must be specified at each cell in the grid, the total number of variables in such a model is of the order 10 million. In other words, you wouldn't want to run this on your desktop. ECMWF has some of the world's fastest computers tucked away in its basement in Reading, a half-hour train ride from London. The Japanese Earth Simulator, with its 5,000 processors, is even zippier and can perform trillions of calculations per second—making it about a million times faster than the ENIAC computer of 1950.[26]

Observations are carried out on board a range of platforms— weather balloons, ground stations, commercial airliners, boats, satellites. Part of the joy of meteorology is in the hobby-shop nature of much of the equipment. I once attended a meeting in California about the latest observation techniques. In the first talk, a meteorologist from a private company demonstrated his device, called a radio-sonde, for measuring pressure and temperature at high altitudes. A tube a couple of feet long, it was packed full of electronics, including a transmitter to beam the information to a weather station. Such devices, which are in common use by weather centres, are dropped from airplanes. A little parachute opens, and they float gently to the ground, recording the weather as they go.

The next offering was a balloon that did much the same thing, except that it was launched from the ground and floated around with the wind. But both the balloon and the parachute were totally outclassed by a company that had developed a radio-controlled plane that had flown all the way from the U.S. to Europe—a world record for a pilot-less device. They were working on a version that could float around in the atmosphere indefinitely, taking its energy from the sun while it continuously recorded the weather.

This was hard to beat, but there was one challenger: a group of Canadians from Vancouver who had come up with a way to do atmospheric measurements over the ocean. Less data collection takes place over the oceans, which for Western Canada (among other places) presents a problem, since the weather there tends to come from the west (i.e., over the ocean). So the idea was to float a buoy that was armed to the teeth with clusters of rockets, like a hedgehog. Every half day or so, one of the rockets would shoot up into the sky. At the top of its trajectory, a little parachute would again open and it would float down, taking measurements as it dropped.

The Canadians were working on some technical problems. One was how to stop seagulls from sitting on the rockets. Another was to make sure that the rockets were aimed straight up, even in a storm, to prevent the possibility of accidentally firing at passing ships and starting a war.

The ultimate piece of kit, of course, is the weather satellite, like the Geostationary Operational Environmental Satellites (GOES), which hover 35,800 kilometres above the earth in geo-synchronous orbits.[27] The planet is now constantly monitored from above by cameras and sensors, like a dodgy customer at a security-conscious bank.

Once measurements are obtained, by whichever means, they are transmitted to weather centres. Since the observations are made at a mix of locations around the planet, they must first be interpolated in some way to get the values at the grid points, a procedure known as data assimilation. Every measurement is subject to error, so the assimilation scheme attempts to smooth or filter spurious results. In modern schemes, this is done by adjusting the observations so they are roughly compatible with model predictions from several hours before. The measured state of the atmosphere is called the analysis, and that forms the initial condition for the model forecast. Every few hours, this is passed to the GCM, which cranks out the latest forecast. If there is a modern equivalent of the oracles of ancient Greece, they reside somewhere in the circuits of these computers. Every day they are fed data, and every day they are consulted for a prediction of the future.

But the predictions still need to be interpreted, just as at Delphi. This is done by human forecasters, often working for private companies, who apply their knowledge of local conditions to improve the result.[28] Weather prediction has grown into a multi-billion-dollar industry; forecasts are routinely supplied to all manner of businesses affected by the weather, such as agriculture, energy, transport, and retail. Consumption of soft drinks, ice cream, movies, medicines, and many other goods change with the weather. Power companies use forecasts to estimate demand and make the expensive decision of whether or not to bring on additional generators. Finance companies offer contracts known as weather derivatives, which insure against anything from a wet winter to the number of frost-free days at an airport, with a worldwide market estimated at $5 billion.[29] When the oracle speaks, they listen.

MEASURING ERROR

So how reliable are these oracles? To measure the error, we must first choose a metric, a measuring stick. In the Lorenz system, errors were expressed as distance in three-dimensional space. In this Euclidean metric, as it is known, each variable carries the same weight. A GCM, however, contains variables of different types (such as pressure, temperature, and so on) at different locations. Since pressure is not directly comparable to temperature or to windspeed—for one thing, they are in different units—we either have to translate between them

or come up with some partial measure. A popular choice in meteorology has been the 500 mb (millibar) height, which is not a pressure
but the height at which the atmosphere has a pressure equal to 500
mb. Pressure decreases with height, as Blaise Pascal showed, and 500
mb is about half the pressure at ground level. This metric is often
plotted as a kind of contour map: the hilltops represent areas where
pressure is higher than usual and the valleys represent low pressure.
Meteorologists can read such maps and pick out features that indicate
large-scale weather systems, though the relationship with weather on
the ground requires much interpretation.

To obtain a single number—a kind of distance between the
forecast and the observations—we can take the RMS difference
between the two over a selection of grid points. This gives a sense of
the average error over that area. The principle is the same as that for
the Lorenz system, but the errors are summed over some thousands
of grid points, instead of just three variables. Different metrics, in
different atmospheric variables, give different results. A more complete, but harder to compute, metric is known as total energy.[30] It
translates the different quantities, such as windspeed, temperature,
and pressure, into compatible units of metres per second and measures them globally at all grid points. The advantage of total energy
is that it accounts for all sources of error.

Finally, we need some benchmark for weather models to beat.
Meteorologists typically use one of the so-called naïve forecasts. The
first of these is to say that the weather will be the same as the climatology—that is, the average for that day of the year. In the shift map
of Chapter 3, this was the "middle of the dough" forecast. The second naïve forecast is persistence, which says that the weather tomorrow or next week will be the same as today's. If the forecast beats one
or both of these, on average, by even a negligible amount, then it is
said to have "skill." Of course, the word here does not have its usual

meaning, because neither climatology nor persistence are very good, but skill does imply that the model is on the right track.

Given these provisos, we can estimate model accuracy. GCMs have now been around for about fifty years, and over that time, they have improved in a slow, incremental fashion. The greatest improvements, and the most commonly cited statistics, have been in the 500 mb metric, which now show skill, compared with persistence or climatology, for up to about a week.[31] Some days the models will do better, other days worse, but over many separate trials, they will have a slight edge. However, the 500 mb height is not the same as the weather—you can't feel it on your face. It corresponds to an average altitude of 5.5 kilometres, which is around where pilots like to fly exactly because it is "above the weather."[32] It is also an intrinsically low-error metric, since it is measured away from most turbulence and storms and tends to average out errors at lower levels.

Predictions of temperature and windspeed closer to the ground are more difficult. For specialized sectors, such as wind farms, where a small statistical edge in windspeed prediction can translate into economic value, it can be worth using forecasts out to five days.[33] For most situations, though, forecasts are noticeably useful for only two or three days, with errors increasing rapidly over that period and then growing more slowly.[34] Beyond that, the climatological forecast works about as well—but it doesn't require a GCM or a supercomputer, or even any interpretation. The most difficult feature to predict (and one of the most important) is precipitation, which has been more resistant to improvement and shows little skill past twenty-four hours.[35] In fact, forecasters interpreting the models do not, as a matter of principle, take the numerical output literally, but instead use their own subjective knowledge to significantly improve the result (which is why skilled human forecasters have yet to be replaced by machines).[36] Improvements in numerical weather prediction have therefore lagged far behind advances in computer speed and observation technology, despite the great economic value of accurate forecasts.

Of course, certain features of the atmosphere have a time scale of days to evolve and dissipate, so it is possible to detect them in advance and anticipate their direction. An example is hurricanes (or typhoons, depending on the part of the world), which can be the size of Texas. These systems are like massive heat engines that take energy from warm oceans and release it through condensation at high altitudes. The direction of the circulation is set by the Coriolis effect, so it depends on the hemisphere (counter-clockwise in the northern hemisphere, clockwise in the southern hemisphere). Forecasters can now predict the track of hurricanes with an average three-day error of about 240 kilometres, which is useful for people trying to get out of the way.[37] However, estimates of storm intensity are much more difficult and have made little improvement over the past fifteen years.[38] Hurricanes draw their energy from the heat in the top few metres of ocean, and as the winds stir up the water, they may bring cooler water up from the depths. A hurricane's intensity depends on the details of this ocean/atmosphere interaction, which is extremely hard to model or predict. The average intensity may increase if oceans warm because of climate change, which is a concern since the damage caused varies roughly with the cube of windspeed.[39]

Longer-term models, which attempt to predict weather phenomena weeks, months, or even years in advance, are naturally even more prone to error. A number of models have been developed to predict the occurrence of El Niño events, which are caused by upwellings of cool water in the Pacific and are second only to the change of seasons in their effect on global weather. The ocean sloshes around on slow time scales, and El Niño events recur every two to seven years. Severe events can significantly disrupt the global economy: the one in 1997–98 destroyed property in California, caused fires in the Amazonian rainforest, hammered the Colombian coffee harvest, and rang the world climate system like a giant bell. The total damage was estimated at $25 billion (U.S.), though this was offset in certain regions by other benefits, like a warmer winter.[40] The phenomenon is clearly worth predicting—accurate forecasts of, say, temperature trends would be worth great sums to power companies alone. Because El Niño is driven by slowly varying ocean effects, there appears to be a degree of predictability. However, according to one report, comparisons of model results with observations have shown that there "wasn't much skill, if any" over simple estimates. Furthermore, "the use of more complex, physically realistic dynamical models does not automatically provide more reliable forecasts."[41]

Even if it is not possible to make accurate "point predictions" about the weather, it might still be possible to make a probabilistic forecast. Many local forecasters now say that there will be an 80 percent chance of rain tomorrow, instead of coming right out and forecasting rain. One way to go about this numerically is to make an ensemble of forecasts from perturbed initial conditions. If they all call for rain, then confidence is high that rain will occur, while if some call for rain and some for sun, the outlook is mixed. There are two difficulties with this approach, however. First, the model initial condition consists of 10 million or so variables, and there are many different ways to perturb them. Ensemble forecasters typically choose the perturbations that give the maximum effect, which may be highly atypical and potentially says more about model peculiarities than about the weather itself. Second, this only works if you believe that most error is due to initial condition, which is not very likely.[42]

Some modern ensemble schemes also perturb the model by

varying parameters in a random way, adding random terms, or simply using a number of different models. Here it is far harder to choose appropriate perturbations because they occur not just at time zero, as with the initial condition, but must change with time. And since the model equations do not capture every aspect of the system, there is no reason to believe that any settings of the parameters will reproduce the actual weather. Also, to make a true probabilistic forecast, we need to know the probability that a parameter will vary in a certain way, which is awkward if a correct value does not exist.[43] The results from ensemble experiments are hard to interpret—the area has evolved a rich and forbidding array of statistical tools—but these schemes appear to offer only slight improvement over simply running a single high-resolution forecast.[44]

If weather prediction is so hard, you may wonder how publications such as the *Farmer's Almanac* can claim to make reliable forecasts months in advance, based on things like sunspots, the planets, and the tidal action of the moon. The answer is that they don't do any better than climatology.[45] That hasn't stopped almanacs from being consistent best-sellers for about the past 3,000 years. At least they are an amusing read. In the 1700s, Benjamin Franklin wrote his own very popular almanac under the name Richard Saunders:

Courteous Reader,

This is the 15th Time I have entertain'd thee with my annual Productions; I hope to thy Profit as well as mine. For besides the astronomical Calculations, and other Things usually contain'd in Almanacks, which have their daily Use indeed while the Year continues, but then become of no Value, I have constantly interspers'd moral Sentences, prudent Maxims, and wise Sayings. . . . If I now and then insert a Joke or two, that seem to have little in them, my Apology is, that such may have their Use, since perhaps for their Sake light airy Minds peruse the rest, and so are struck by somewhat of more Weight and Moment.[46]

A curious feature of prediction of any sort is that there is little correlation between the forecast accuracy and the amount that people are willing to pay. An acquaintance who used to work at Enron told me that some power companies even paid for forecasts a year in advance. (It would have been cheaper to buy the almanac.) As Cato the Censor said of ancient Rome, "I wonder how one augur can keep from laughing when he passes another."[47]

So what's going on? Can chaos be responsible for all this error? Or could the problem lie deeper?

COMPLICATIONS

According to that great savant, U.S. Secretary of Defense Donald Rumsfeld, errors can arise because of both the known unknowns and the unknown unknowns.[48] In weather forecasting, the known unknowns are the errors in the initial condition. We know that a particular cell in a grid has a certain average temperature, pressure, and so on, but we also know that our estimate of those quantities—the analysis—is in error. The measurement devices are accurate only to a certain tolerance; the temperature, or any other quantity, may vary widely over a single cell; and finally, there are many areas of the globe, like the oceans, where measurements are quite sparse. In these areas, the data-assimilation scheme must interpolate from the few measurements that exist, a procedure that is prone to error. However, we at least know that the errors exist, and therefore can estimate their magnitude.

The second source of errors is the unknown unknowns—the things we don't know that we don't know. The actual weather prediction is carried out by a GCM that's based in part on physical laws, such as conservation of energy. However, it still has to make a large number of approximations. For example, the coarse resolution means that the weather over a single grid cell is averaged out, and anything smaller than a cell, such as a single cloud, a small storm, or a detail of a coastline, doesn't appear. Turbulent eddies in the atmosphere have the effect of transferring energy from large-scale (kilometre-size) to small-scale flows.[49] As Lewis Fry Richardson put it, paraphrasing Jonathan Swift, "Big whirls have little whirls that feed on their velocity, and little whirls have lesser whirls and so on to viscosity—in the molecular sense." Models resolve only big whirls, and must estimate the effect of small whirls. Experiments using cellular automata, which simulate idealized local interactions between parcels of air, show that turbulent flow is not generally sensitive to initial condition, but it still eludes computation using equations.[50] The atmospheric motion near the ground is particularly difficult to model, since the flow interacts with the planet's surface, is highly turbulent, and must take into account local heat fluxes (which depend on factors such as soil type, vegetation coverage, and so on).

The biggest contribution to error, though, is water. From a dynamical point of view, the most important substance in the atmosphere is not air but the water that air contains. There isn't that much of it: if all the water in the sky were to fall to the ground as rain, it would form a layer about an inch high.[51] But the high thermal capacity of water means that it can contain more than its fair share of energy in the form of heat. Indeed, the top few feet of water in the earth's oceans contain as much heat as the whole atmosphere (and this drives El Niño). The release of latent heat energy, which occurs when water vapour condenses into clouds, is one of the major drivers behind atmospheric circulation. Clouds in turn reflect solar radiation, cause precipitation, and determine much of what we call the weather. These processes are poorly understood and cannot be modelled from first principles, so modellers are forced to use parameterizations: ad hoc formulas that attempt to fit the available data.

Take clouds. In a GCM, one cell of the grid might be assigned a certain degree of "cloudiness" to reflect the average cloud cover. Variables such as temperature or pressure are by definition averages over a large number of molecules, so it makes sense to average them over a cell. Clouds, however, are entities with a high degree of structure. Leonardo found that a human body would not fit easily into a square, and clouds do not fit into a Cartesian box.

A common game for children is to look for clouds that resemble animals, and it is perhaps fitting that the first system for categorizing clouds was invented in 1801 by the French botanist and zoologist Jean Lamarck, whose theory of evolution preceded Darwin's. The system that was eventually adopted worldwide was proposed independently in 1803 by the Englishman Luke Howard. It divided clouds into three basic types: cirrus, cumulus, and stratus. Cirrus clouds are high, wispy clouds that consist almost entirely of ice crystals, cumulus clouds are mid-level, and stratus are the closest to the ground. There are numerous subcategories, from the cauliflower-like cumulus congestus to the threatening, anvil-shaped cumulonimbus. Clouds also come in a huge range of sizes and have a fractal, scale-free property: wisps of cloud viewed up close from an airplane window can look similar to massive cloud systems in satellite pictures. You can't talk about an average cloud, or even an average-size cloud, in the same way you can talk about an average molecule of air.

The formation and dissipation of clouds is also a complex dynamical process. Clouds are a mixture of minute water droplets and ice crystals. Their growth depends on a wide variety of factors, such as humidity, temperature, and the presence of small particles in the air. These particles—which include dust, smoke, sea salt, droplets of sulphuric acid and ammonium sulphate caused by algae emissions,[52] plant fragments, industrial chemicals, spores, pollen, and even human dandruff[53]—act as nuclei on which water can condense. (Dubious attempts at weather control often involve shooting such particles into the atmosphere to "seed" clouds.) Because of the intrinsically social nature of water molecules, the rate of condensation on a droplet depends on the size and, particularly, the radius of the cluster already there. Like people choosing a restaurant, water molecules prefer a spot that is already well attended. As a result of this positive feedback, the growth of each droplet depends in a non-linear way on its size; in any given cloud, there will be a range of droplet sizes.

Trying to model a cloud is about as easy as trying to hold one in your hands. Despite the heroic efforts of meteorologists, the best that a GCM can do is to assign rough values for cloud properties for each cell, making them vary in some plausible way to account for things like temperature and humidity. Such parameterizations may be based to a degree on physics, but they are a long way from Newton's laws of motion. (There may be a law of gravity, but there isn't a law of clouds.) They are a major source of error, especially since estimates of cloud cover affect the calculations of temperature, humidity, and so on, which are the calculations used to make the estimates in the first place. To predict how clouds change and evolve with time, details matter. And because clouds exist over a huge range of scales, there is no particular grid size that is small enough to capture all the information. Even if the resolution is improved, new parameterizations will be needed to model the fine-scale physics. The number of model variables will therefore explode, and forecast accuracy may actually get worse.[54]

In many ways, it makes more sense to view clouds as emergent properties of a complex system, rather than as something that can be computed from first principles. In a paper about human cognition, psychologists Esther Thelen and Linda Smith compare the formation of clouds and storms to the development of cognition:

> There is a clear order and directionality to the way thunder-heads emerge over time. . . . But there is no design written anywhere in a cloud. . . . There is no set of instructions that causes a cloud . . . to change form in a particular way. There are only a number of complex physical . . . systems interacting over time, such that the precise nature of their interactions leads inevitably to a thunderhead. . . . We suggest that action and cognition are also emergent and not designed.[55]

As will be discussed further in Chapter 7, there are countless other sources of error in the prediction of weather, many of which relate to the complex positive and negative feedback loops that regulate the climate. In fact, while the GCM is a huge and sophisticated project—perhaps the most complex mathematical model yet produced—it is still only a crude approximation of the ocean-atmosphere system.

GETTING THE DRIFT

While errors in the initial condition can grow exponentially (which is not the same as rapidly) in chaotic systems, the errors in the model are even harder to predict because they grow in a cumulative and dynamic manner. The model error at the initial time is different from the model error a while later. And since the model is an ODE, which specifies the rate of change of variables with time, the errors are expressed not in terms of atmospheric quantities like temperature and pressure, but in terms of their rate of change. In the Pythagorean scheme, they are not at rest, but in motion.

When I first started researching model error as a Ph.D. student in 1999, the consensus opinion among experts was that it was a small effect compared with chaos. As the story went, "In the early years of NWP [Numerical Weather Prediction], forecast errors due to simplified model formulations dominated the total error growth. . . . By now, however, models have become much more sophisticated and it is the errors that arise due to [chaotic] instabilities in the atmosphere (even in case of small initial errors) that dominate forecast errors."[56] Some even thought model error wasn't worth investigating. An anonymous reviewer of a paper said that "many people at the operational [weather] centers are not convinced that the effect of model error on forecasting is such that it warrants a great deal of research effort."[57]

As a "mature" student who had already worked several years on the design and testing of magnet systems for particle accelerators, I found this position a little strange. As any engineer knows, there can be a big difference between theory and reality. Even if you have complete knowledge and control of the materials used in construction, they can still behave in unexpected ways when they are all put together in a structure. This is what happened with London's Millennium Bridge, better known as the wobbly bridge because of the way it started to weave and shake the first day a crowd walked over it. An engineer from the firm that designed it told me they had done three detailed analyses of its motion, but none had revealed a potential problem. Perhaps the reason engineers are more concerned with model error than forecasters is because the latter never get sued (though people have tried).[58]

My Ph.D. supervisor was Lenny Smith from Oxford University (now also at the London School of Economics), who is an expert on chaos theory and non-linear systems, as well as an inspiring mentor.[59] He suggested model error as a topic for the very reason that it would be new, uncharted territory. There were only a handful of experiments that compared the predictions of two different models—say, an American one and a European one—with the actual weather.

These experiments showed that the difference between the model and the weather was similar for either model, so it was assumed that the sensitivity of forecasts to the choice of model was small. However, the fact that two models are wrong by a similar amount does not mean they are both right: they may just be wrong in similar ways. Also, while Americans and Europeans are dissimilar, their models are quite alike. They are each written by meteorologists who read the same textbooks and attend the same conferences and incorporate one another's improvements.

The easiest way to measure the difference between two models is to compare them to each other. In collaboration with ECMWF, we set out to compare a low-resolution model (which would have been the best available just a few years earlier) and a more recent high-resolution model. The latter would play the role of "truth"—the real weather—while the former would be the "model," with the error being the difference between the two. The idea was to see how close the model versions were to each other: if errors were large, it implied that the models had not converged. Later, we would compare the operational model, used in everyday forecasting, to actual observations of the weather. Neither of these basic reality checks had ever been performed before.

Model error was measured using two methods. The first was to look for shadows. If error is a result of chaos rather than a problem with the model, it should be possible to perturb the initial condition slightly so that the resulting model trajectory, known as a shadow, stays close to truth.[60] On the other hand, if sensitivity to initial condition is low compared with model error, then no small adjustment will help. It is like an archer trying to hit a target in a strong wind. Shadows have a fairly long history in the field of non-linear dynamics, but the technique had not been used with weather models.[61] The models have millions of variables, and a perturbation to the initial condition can be made to any or all of them. Computing shadows

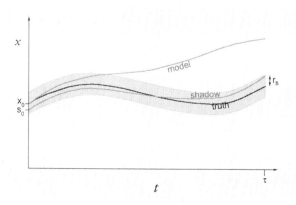

therefore involves a search over many possible choices, but it can be done using a suitable optimization program.

FIGURE 4.7. Simplified schematic diagram of a shadow orbit. When the model is initiated at the "true" initial condition, it soon exceeds the prescribed tolerance (r_s). The shadow trajectory starts at a perturbed initial condition and shadows (stays within the tolerance) for a longer time.

The second method, calculating the model drift, is a simple way of estimating model error, but it has the benefit of being easy and fast to apply. The calculation is performed by making a large number of short forecasts at regular intervals—say, every six hours—along the true trajectory. The forecast errors are then summed together, and the magnitude of the sum is the drift. Because the forecasts are constantly being set back to truth, this filters out the effects of chaos, so the drift approximates the error resulting from the model alone.[62] It adds up all the small, moment-by-moment errors that push the model away from the truth and can be used to estimate the expected shadow performance.

Both the drift and the shadow calculations were in agreement. They showed that most error was the result of differences in the models. Since we were comparing two models of different resolution, this meant that they had not converged on a single version of reality.[63]

CURVING UP OR CURVING DOWN

Of course, what we really cared about was how the operational forecast model, used every day to make predictions, compared to the analysis. The results here were even more striking. A drift calculation showed that nearly all the error over the first three days was the result of the model. Observational errors obviously also played a role; however, this was reduced in part because of the way the data assimilation was being performed. Chaos appeared to cause only a slight amplification of the model errors. Even then, the shape of the error curve showed no sign of the exponential growth that we would expect to see in a strongly chaotic system. In fact, over the first couple of days, it grew roughly with the square root of time, so by day two the error was about twice what it was at twelve hours (as seen in figure 4.8 on page 164).[64]

The reason, we realized, was that the model errors were behaving like a random walk. This term, commonly used in economics to describe price fluctuations caused by external shocks, was first introduced in a 1905 paper in the journal *Nature.* The paper aimed to determine how far a drunken man walking randomly in an open field could be expected to travel.[65] The man takes a step in a random direction, then another step in a random direction, and so on, gradually getting farther and farther away from his starting point (which is bad if he is looking for his keys). The expected distance travelled grows with the square root of time. In weather forecasting, the model was like a sober person marching in a straight line, while the weather was the drunken friend weaving randomly from side to side and getting gradually farther away.

In many ways, square-root error growth is the opposite of the exponential growth in figure 4.5. The former grows most quickly for short times, the latter for long times. The former curves down with time, while the latter curves up. It is impossible to confuse the two. So when I presented the results in a seminar at ECMWF, I was surprised when I put up a plot of error growth, and someone

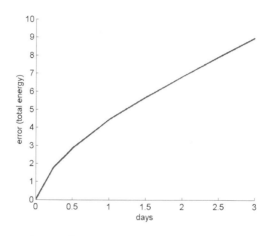

interrupted my talk to say that the plot must be wrong, since error growth has positive curvature, not negative. Others in the audience murmured in agreement.

FIGURE 4.8. Plot of error growth in total energy for weather model versus analysis. Errors grow roughly with the square root of time over the first couple of days.

I was naturally disconcerted, and though I continued with my talk, I wondered if I had made a mistake in the calculations I had performed to produce the figures. My apprehension increased the next day, when I received an e-mail from one of the top research heads at ECMWF, which said that he had checked a plot of wind errors, and in stark contrast to mine, his plot "certainly [did] show positive curvature."[66] We therefore decided that someone there should try to reproduce my results by plotting the errors as a function of time. Either they would curve up, or they would curve down.

That weekend, I nervously looked through old ECMWF reports showing plots of error growth. Although none of the experiments had been performed in a global metric like total energy, and all were only incomplete snapshots of the errors, they still seemed in agreement with my results, and there was no sign of exponential growth. Also, I noticed that the authors of the reports, almost in an act of self-censorship, had consistently omitted the first couple of time points, which made the negative curvature less obvious. But surely ECMWF, which had perhaps the best weather models in the world, would know what error plots looked like.

The next week, I received an e-mail containing the recalculated error results. They were exactly the same as the ones I had presented. So it hadn't been some numerical mistake on my part—the errors really did grow with negative curvature. It seemed that the desire to blame chaos had affected people's basic shape-recognition skills. I thought this would settle the matter, but it turned out to be only the beginning. The weather centres refused to accept that the error could be caused by the model, and they advanced a number of theories (but no more experiments) about how the results were compatible with exponential growth.

IRRATIONAL NUMBERS

One of the more popular theories was that the doubling time of the exponential growth changed, so it was fastest at short times, then progressively slowed. A couple of months later, a workshop on atmospheric predictability was held at the Naval Research Laboratory in Monterey, California. During one session, the recalculated error plot was put up to determine the doubling time. As I watched, several representatives from the top weather centres of the United States and Europe formed a "consensus" that the doubling time must be about a day or two, because it is faster than a day at the beginning and more than that at the end. I protested that the graph was not an exponential curve, so it did not have a doubling time. The errors grew with the square root of time. But it was like Hippasus telling the Pythagoreans, who believed only in rational numbers, about the existence of the square root of two. I was lucky they didn't take me out to sea and throw me overboard, as the Pythagoreans were rumoured to have done. The published meeting report stated that though discussion was "very animated," the conclusion was that "predictability limitations are not an artifact of the numerical model."[67] So according to the "consensus," the model was effectively flawless.

This all reminded me of the Greeks' attempts to replicate the motion of the planet by incorporating more and more epicycles into their model. The "consensus" opinion could perhaps explain why the error curves didn't look exponential, but it would have been a coincidence that they varied so closely with the square root of time. According to Ockham's Razor, the simplest explanation is usually the best, and in this case, the simplest was that the error was caused by the model.

The more closely we look at the assumption that chaos lies behind forecast error, the weaker it seems. Even the techniques used to measure the doubling time of errors appear to have been chosen to produce a fast result. The usual method is to make small perturbations to the initial condition and see how the error grows (as in figure 4.5 for the Lorenz system). Experiments with weather models, however, were performed using a technique known as lagged forecasts, which appeared to give doubling times as fast as a day.[68] It turned out that the rapid growth was a result of the special type of perturbation used, which was large in all variables except those being measured. As the error propagated from the other variables, it appeared to grow rapidly. But when the experiment was repeated in a global metric, which took all errors into account, the rapid growth disappeared.[69]

The way the analysis is typically prepared is also questionable. In 1941, the Nazi meteorologist Franz Baur told Hitler that the upcoming winter in the Soviet Union would be mild or normal. Hitler decided to launch an invasion, but his unprepared army was frozen in its tracks by a winter of almost unprecedented severity. When informed of the conditions, Baur said, "The observations must be wrong."[70] A similarly confident attitude is adopted in the analysis procedure, which adjusts atmospheric observations to better fit the model predictions from a few hours before.[71] In other words, the model is favoured over reality. The practice is defensible to a point, because it helps smooth the observations. However, errors are then measured against the analysis rather than the original observations. Because the analysis is made to be more like the model, this reduces published forecast errors, therefore demonstrating the accuracy of the model. The GCM has become the embodiment of the Pythagorean ideal of an ordered, rational universe, and weather research centres often seem concerned more with protecting it than with making accurate predictions of the future.

BAD WEATHER

In January 2003, the American Meteorological Society (AMS) held its annual meeting at the downtown convention centre in Seattle, where I then lived. Thousands of meteorologists descended on the city from research centres all over the world, clutching laptops, briefcases, and tubes containing poster presentations.

Scanning the local newspaper, I saw the meeting announced with a quote, supplied by the AMS media advisory, which claimed that "numerical weather prediction is widely regarded to be among the foremost scientific accomplishments of the 20th century." On the back page was the weather forecast. Typically for Seattle in January, it appeared to consist of random selections from the words "rain," "drizzle," "cloudy spells," and "sun breaks."

To most people, our less-than-awesome ability to predict the weather doesn't quite rank with, say, putting a man on the moon or discovering DNA. It is true that enormous progress has been made in understanding how the weather works, which is laudable, but despite massive increases in research effort and computer power, this progress has not carried over into accurate predictions. If anything, weather forecasting has been one of the great underachievements in science; there's a huge gulf between what seemed attainable early in the last century and what is possible today. This reflects not on the quality of the scientists—many brilliant people have worked or are working in the area—but on the complexity of the problem.

It is nonetheless remarkable that even though GCMs had been around for fifty years, the topic of model error was still viewed as "new, uncharted territory." When I first performed a search to see what else had been written on the subject, I found that only a handful of research papers contained the words "model error" in the abstract, and none of them had made a serious, concerted effort to measure it.[72] So why was there so little work being done in this area, and why were meteorologists so eager to adopt chaos as the cause of forecast error?

Part of the reason is institutional. Aristotle once said that man is a political animal, and scientists are no exception. As the engineer Bart Kosko puts it: "Career science, like career politics, depends as much on career maneuvering, posturing, and politics as it depends on research and the pursuit of truth. . . . Politics lies behind literature citations and omissions, academic promotions, government appointments, contract and grant awards . . . and most of all, where the political currents focus into a laserlike beam, in the peer-review process of technical journal articles—when the office door closes and the lone anonymous scientist reads and ranks his competitors' work."[73] It is hard to publish papers that demonstrate that models don't work, or that expensive strategies are poorly conceived—especially when the people reviewing the manuscripts are the ones who developed the models and strategies in the first place. The peer-review process has many benefits, but it can easily slide into a kind of self-censoring avoidance mechanism.

Another problem is that scientists, like anyone else, become attached to fashions and fads. In the 1980s, as Stephen Wolfram observed, signs of chaos were detected not just in weather forecasting but "in all sorts of mechanical, electrical, fluid and other systems, and there emerged a widespread conviction that such chaos must be the source of all important randomness in nature."[74] But while some mathematical equations show sensitivity to initial condition, "none of those typically investigated have any close connection to realistic descriptions of fluid flow."[75]

Perhaps a deeper problem, though, is that investigating model error—calculating the drift, finding shadow orbits—is like exploring the shadow side of science. Instead of saying what we can do, trumpeting our superior intelligence over brute nature, it shows what we cannot do. It demonstrates ignorance instead of knowledge, mystery instead of clarity, darkness instead of light. By embracing chaos as the cause of forecast error, meteorologists could maintain the illusion that models were essentially perfect.[76] The same temptation exists in other branches of science. As we'll see in Chapter 6, economists managed a similar trick in the 1960s, when they blamed unpredictability on random events external to the economy.

Of course, the fact that model error is large now does not imply that the models are in some obvious way flawed, or that they will not continue to improve. The models do an excellent job of capturing the atmospheric dynamics, to the extent that this is achievable using equations. Further advances and adjustments, coupled with better methods to measure model error, will yield continued improvements. But increasing the resolution of the models offers diminishing returns.[77] GCMs will be better in a hundred years' time, but there is no reason to anticipate the arrival of perfect one-week weather forecasts.

Meteorology's greatest contribution has probably been to provide warnings of storms or other short-term phenomena, or more recently of fragilities in the climate system that we may inadvertently be affecting. Atmospheric scientists have often led the way in pointing out the dangers of human impacts on the environment. We would have had no idea about the decline in the ozone layer or the rise in atmospheric carbon dioxide if no one had gone out and measured them, little idea of their importance to the planet if no one had modelled it. For such purposes, simple models that do not attempt to capture the full system in all its glory are often effective. (The danger of global warming, for example, was first identified by the Swedish physical chemist Svante Arrhenius over a hundred years ago.) The developing area of complex systems research may not directly improve weather predictions, but it will build on our understanding by offering new conceptual approaches.

SUN SCREEN

Human effects on the environment have a way of coming back at us. A famous example was the discovery that artificially synthesized chemical compounds known as chlorofluorocarbons, or CFCs, were destroying the ozone layer. CFCs were introduced in the 1930s as non-toxic cooling agents for refrigerators and air-conditioning units. In the middle part of the twentieth century, air-conditioning grew enormously in popularity. This increasing demand led to the mass-production of CFCs, and their slow release into the environment.

In the 1970s, the British scientist James Lovelock used his new invention, the electron-capture gas chromatograph, to detect the presence of aerosols in the atmosphere. He discovered that CFCs were found at surprisingly high concentrations. The chemists Sherwood Rowland and Mario Molina came to the conclusion that CFCs could degrade ozone (a form of oxygen) in the upper atmosphere. This effect was alarming because the ozone layer, as it is known, acts as a protective screen against DNA-destroying ultraviolet radiation. The ozone had accumulated in the atmosphere as a by-product of billions of years of photosynthesis by ocean-living organisms, but CFCs were eroding it in mere decades. Chlorofluorocarbons are also potent greenhouse gases.

Rowland and Molina were at first ignored by the companies responsible for making the compounds, but their concerns were borne out by the discovery of an "ozone hole" above Antarctica. Helped along by a degree of media excitement, this led to the banning of CFCs. Even if atmospheric modelling can't predict the future, it can sometimes help avert disaster.

We can summarize with the following points, many of which also apply in modified form to the other systems studied in this book:

- **The ocean/atmosphere system is complex and based on local interactions.** Structure exists over all scales, from microscopic to global.
- **Features such as clouds or storms can be viewed as emergent properties.** Their behaviour cannot be computed from first principles.
- **GCMs use parameterizations to approximate these properties.** The difference between the equations and reality results in model errors that limit prediction.
- **Increasing model complexity does not necessarily lead to error reduction.** More parameters need to be approximated. As a result, improvements in prediction accuracy, especially for key metrics such as precipitation, have lagged far behind improvements in computers, observation systems, and scientific effort.
- **It is still possible to make general warnings.** These can often be made with simple models that do not attempt a detailed simulation of the entire climate system.

If all that was at stake was the short-term weather, then the longer-term accuracy of weather models would be of no great concern, and the cause of error would be academic. The problem, however, is that if forecasts for next week are wrong because of seemingly unresolvable model errors, we have no reason to think that we'll be able to predict, decades in advance, the even more complex effects of global warming on the biosphere.

This assertion may appear to be completely negative. Perhaps paradoxically, though, I believe the opposite to be true. The realization that the earth system is inherently unpredictable—coupled with a deeper understanding of and respect for complex systems—may turn out to be highly liberating.

In the next chapter, we look at a more personal kind of prediction: that of the destiny that is written in our genes. Just as weather prediction is "only" fluid flow, the reading of genetic information is "only" biochemistry. Again, though, we will see that something seems to get lost in translation.

5 ► It's in the Genes
Predicting Our Health

JODY: Scorpion wants to cross a river, but he can't swim. Goes to the frog, who can, and asks for a ride. Frog says, "If I give you a ride on my back, you'll go and sting me." Scorpion replies, "It would not be in my interest to sting you since as I'll be on your back we both would drown." Frog thinks about this logic for a while and accepts the deal. Takes the scorpion on his back. Braves the waters. Halfway over feels a burning spear in his side and realizes the scorpion has stung him after all. And as they both sink beneath the waves the frog cries out, "Why did you sting me, Mr. Scorpion, for now we both will drown?" Scorpion replies, "I can't help it, it's in my nature."

FERGUS: So what's that supposed to mean?

JODY: Means what it says. The scorpion does what is in his nature.

—Neil Jordan, *The Crying Game*

. . . CGCGGTGCTCACGCCAGTAATCCCAACACTT . . .

—Sequence of human DNA (basepairs 100000–100030 of chromosome 16), from the Human Genome Project[1]

GOOD BREEDING

To prophets of different persuasions, anything from tea leaves to animal entrails to segments of the Bible can form a kind of text that can be probed for glimpses of the future. In Kepler's day, a child's destiny was thought to be determined—or at least strongly influenced—by the positions of the sun, the moon, the planets, and the stars at the time of birth. New parents could pay an astrologer (such as Kepler or Tycho) for a detailed horoscope and get the weather for the year thrown in as well. Now, we have the option of consulting medical experts, who scan stretches of DNA instead of the heavens for hidden portents. With procedures such as amniocentesis, an expectant mother can have her unborn fetus tested for a range of genetic diseases. And with the completion of the Human Genome Project, there is the promise that DNA will give up information about a broad range of physical and mental traits. How strong will the child be? How smart? How long-lived? In this chapter, we look at genetic prediction: how scientists use biochemistry and statistics to determine the effect of genes and inheritance on an individual's health.

Humans have always been fascinated by the question of how traits are inherited. Aristotle was of the opinion that semen contained particles of blood, known as pangenes, which were passed on from generation to generation. In *The Eumenides,* the playwright Aeschylus has Apollo say, "The mother is no parent of that which is called her child, but only nurse of the new-planted seed that grows. The parent is he who mounts."[2] A popular theory in the seventeenth century was that the sperm or the egg contained a minuscule homunculus—a tiny version of a human being that grew into an embryo. This theory was stunningly confirmed in 1677, when the Dutch naturalist Antonie van Leeuwenhoek observed sperm under the microscope and saw what he believed was a tiny human being. (Microscopes have improved since then, though we still tend to see what we are looking for.)

The story of modern genetic prediction begins with two men, both born in 1822: the Victorian polymath Sir Francis Galton, and the Austrian monk Gregor Mendel. There are essentially two ways to make scientific predictions. The first is to look for statistical patterns in past data and predict that they will continue. This was the technique used by Sir Gilbert Walker to detect the El Niño pattern (though there, prediction proved more difficult) and is especially popular in finance. The scientist may then work backwards to find a causal explanation, but this is at the risk of imposing a story that seems plausible but is oversimplified or wrong. The second method is to use mathematical models derived from physical principles. Mendel's work was based on studying the simplest traits; although little known or appreciated in its time—Galton learned of it late and paid it no great attention—it eventually led to a kind of physics for life. Galton, who was concerned with complex traits such as human intelligence, was the pioneer of the data-driven approach, and his work is an excellent example of both its uses and its risks.

Among Galton's many contributions to society, which include the newspaper weather map, fingerprinting, and a description of how to make the perfect cup of tea,[3] we can include the coining of the phrase "nature and nurture." He used the phrase in the title of his 1875 book, *English Men of Science: Their Nature and Nurture,* and might have been inspired by Shakespeare's *Tempest,* in which Prospero describes Caliban as "a devil, a born Devil, on whose nature nurture can never stick."

Galton's thesis was that human ability was a function of nature rather than nurture, that it was in the blood. In a two-year experiment in collaboration with his cousin Charles Darwin, he tried to prove this by injecting the blood of lop-eared rabbits into grey rabbits, and vice versa, to see how it affected their progeny.[4] (It didn't.) He did argue that exceptional ability could be inherited by the usual means, however, and demonstrated it by showing that the children of particularly eminent people were more likely to themselves be eminent. (Here, eminence was measured by mention in a biographical handbook, and by whether or not someone's obituary appeared in the *London Times*.[5])

According to Galton, traits such as ability radiate down the family tree like a kind of Newtonian force, diffusing with distance. In the Law of Ancestral Heredity, parents contribute one-half of inherited traits, grandparents one-quarter, great-grandparents one-eighth, and so on, decreasing by a factor of two with each generation. The sum of all contributions from one's predecessors is ½ + ¼ + ⅛ + 1/16 + . . . which adds in the limit to one.[6] Of course, eminence isn't the only thing that gets passed down. Galton also argued that the children of criminals were more likely to themselves show criminal tendencies: like the scorpion in Aesop's fable, they have it in their nature.

To prove his theory, Galton needed a quantitative way to measure people's traits and characteristics. This exposed a problem with the metric: it is hard to measure what goes on in people's minds. He therefore decided to concentrate on external, physical properties, and he set about measuring the dimensions first of English schoolboys and then of the population at large. During the International Health Exhibition in 1884, he set up an anthropometric laboratory and collected statistics such as height, weight, and strength on thousands of people (who seemed to enjoy being measured, and paid three pence each for the privilege).[7]

WHAT'S NORMAL?

Since Galton could not demonstrate using biology that traits were inherited, he instead tried to prove the point with statistics. He performed his analysis using two techniques, correlation and the bell curve, which are still the basis for modern statistical prediction in areas from demographics to the stock market.

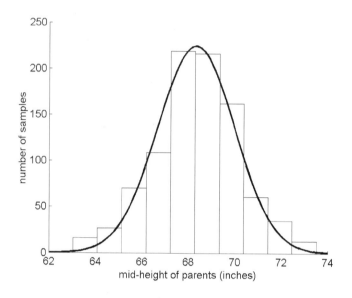

FIGURE 5.1. Histogram of hypothetical height data, based on Galton's report of the adjusted mid-height of parents.[8] The height of each column indicates the number of samples in that bracket. The distribution can be approximated with a bell curve of mean 68.3 and standard deviation 1.65.

First proposed by Abraham de Moivre in 1718, the bell curve, also known as the normal distribution, is a method used to approximate the spread of randomly varying quantities. Laplace used it in 1783 to study measurement errors, and today IQ scores are based on it. Its shape is specified by two numbers—the mean, or average, and the standard deviation, which is a measure of its width. Sixty-eight percent of the results are within one standard deviation of the mean, and 95 percent within two standard deviations. Figure

5.1 shows a histogram of 928 measurements of height (the data, though hypothetical, is similar to that used by Galton). The height of each column indicates the number of samples in that bracket; most are clustered close to the mean, then the numbers tail off in a smooth way. Also shown is a bell curve that fits the data quite well.

There's something almost magical in the way that apparently random data turn out to correspond to a mathematical function. The name "normal distribution," however, is somewhat misleading. Just as most numbers are not rational, so most statistical samples are not normal. The normal distribution owes its popularity to its mathematical properties: in the same way that a linear system is the sum of its parts, the sum of many bell curves, or other distributions, is a bell curve. If the heights of men and women each follow a bell curve, then a couple's adjusted mid-height, which is the average of the parents' heights after adjusting for the difference between men and women, will do the same. As we'll see in the next chapter, many quantities that have been modelled using a normal distribution have turned out not to be so normal after all.

Relationships between different sets of data, such as the heights of parents and children, were quantified by Galton using the concept of correlation. Suppose that one set of data Y is plotted against another set of data X. If the two have the same degree of noise or scatter, and a straight line is drawn which best interpolates the data, then the correlation is defined as the slope of the line (given by the rise over run). This is 0 if the line is horizontal, 1 if the angle is 45 degrees, and −1 if the angle is −45 degrees. A 1 or a −1 means that the data is perfectly correlated. Figure 5.2 (see page 180) shows a scatter plot of children's heights against the adjusted mid-height of parents, based again on Galton's data. Each point represents a particular child/parent data pair. The straight line, determined here by a technique known as linear regression, indicates a positive correlation of 0.63 between the height of parents and that of their offspring: tall parents tend to have tall children.[9]

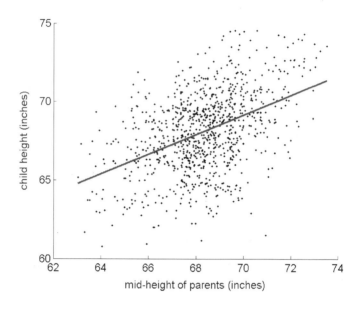

FIGURE 5.2. Scatter plot of children's heights versus the adjusted mid-height of their parents. The slope of the line is 0.63, indicating a positive correlation.

Galton used correlations to argue that traits were passed on from generation to generation, a reasonable assumption to make for height. But it is usually impossible to discern cause and effect from correlation alone, or to know whether both data sets have been influenced by a third factor. The data in figure 5.2 appear to show that tall parents cause tall children, and therefore that height is inherited. However, if the measurements came from poor and rich families—so some were very badly nourished and others well fed—then it could be that height is more a function of diet (nurture rather than nature).[10] Experimenters usually try to screen out such effects (as Galton did, to an extent) by carefully controlling the data.

Since the angle of the line in figure 5.2 is less than 45 degrees (the correlation is less than 1), it means that while tall parents will have taller-than-usual offspring, those children will on average be shorter than the parents—a phenomenon that Galton called regression to the mean. Galton saw this as a kind of negative feedback mechanism that prevents traits from growing without limit. Without regression to the mean, members of successive generations of tall families would grow taller and taller, while short families would shrink to a dot. Regression to the mean implies that variance within a population is bounded, and it's the theory behind a number of slogans still popular in places like Wall Street, such as "What goes up must come down."

If height could be inherited, then so in principle could complex traits like longevity, intelligence, or criminality. In Galton's composite photographs of criminals, he used methods he had "frequently employed with maps and meteorological traces" to pick out correlations in features.[11] A more successful crime-fighting technique was fingerprinting as a means of identification. He developed a classification scheme, which remains in use today, for the intricate whorls and vortices, themselves reminiscent of storm systems on weather maps.[12] Human identity, it seemed, could be mapped and predicted just like the atmosphere.

In the nineteenth century, the heyday of mechanistic science, all of this led to a horrible kind of deterministic logic: if human traits were in the blood, it followed that they could be improved by selective breeding. As Galton wrote, "No one, I think, can doubt . . . that, if talented men were mated with talented women, of the same mental and physical characters as themselves, generation after generation, we might produce a highly-bred human race, with no more tendency to revert to meaner ancestral types than is shown by our long-established breeds of race-horses and fox-hounds."[13] He introduced the word "eugenics" to describe "questions bearing on what is termed in Greek, *eugenes,* namely, good in stock, hereditarily endowed with noble qualities."[14]

The eugenic scheme was perhaps best expressed in Galton's unpublished novel, *Kantsaywhere,* completed shortly before his death. It described a futuristic society whose inhabitants were required to pass a series of exams for health and intelligence. Those who did well were rewarded and encouraged to procreate, while those who were mediocre could have only a limited number of children. The punishment for failure was enforced celibacy. Kepler's parents—a mercenary and an accused witch—might have flunked.

Galton perhaps got the idea from Plato's *Republic.* This too described a Utopian society, where members of the elite Guardian class were to be bred like "dogs for hunting."[15] Their children would be taken to a nursing area, where wet nurses would "provide for their nurture,"[16] while "the offspring of the inferior, or of the better when they chance to be deformed, will be put away in some mysterious, unknown place, as they should be,"[17] so that the Guardians would be kept pure. (As always with Plato, it is hard to tell how literally this should be taken.) More recently, eugenics as a tool for improving society has been supported by people from the Nazis to Winston Churchill, and was part of government policy in places such as Alberta into the 1970s.[18] Communists, on the other hand, championed nurture over nature and believed that traits could be moulded by the state. As Lenin boldly put it: "Man can be corrected. Man can be made what we want him to be."[19] Prediction has always been not just about foretelling the future, but about controlling it. And scientific predictions sometimes say as much about politics and sociology as they do about the system under study.

Smooth or Wrinkly

We now think of ancestral qualities as being passed on through the genes. The idea that traits were transmitted from one generation to the other in discrete form—a kind of quantum theory of heredity— was first proposed by Gregor Mendel. The son of a farmer, Mendel was a keen gardener and a trained scientist (like Galton, he also published in the area of meteorology). In a series of careful experiments, carried out on the grounds of his Augustinian monastery, he tracked thousands of pea plants for several years, pollinating them by hand to control how they mated.[20] He found that certain traits, such as colour and shape, were passed on in a discrete fashion. If a plant whose seeds had a smooth surface was bred with one whose seeds were wrinkly, the resulting seeds would be either smooth or wrinkly, rather than a mix of the two. He proposed that, for any particular trait, each parent plant had two "factors" and would contribute one to the next generation.

Factors could be either dominant or recessive. Suppose that the factor for smooth seeds is S and the factor for wrinkly is W. A child plant inherits one factor from each parent, as shown in table 5.1 (on page 178). The factor S is dominant, so a seed that inherits at least one S will be smooth; a seed is wrinkly only if it inherits two Ws. If the factors S and W are equally distributed in the parents, 75 percent of peas should be smooth and 25 percent wrinkly. Mendel's statistics for his 19,959 plants yielded 14,949 smooth (74.9 percent) and 5,010 wrinkly (25.1 percent), which was Pythagorean in its geometric precision (to the point that statisticians accused him of "correcting" the data). Heredity was all in the factors—or the genes, as they became known at the start of the twentieth century.

PARENT 1	PARENT 2	CHILD
S	*S*	*S*
S	*W*	*S*
W	*S*	*S*
W	*W*	*W*

TABLE 5.1. Each parent contributes one of its factors to the child. *S* denotes smooth, *W* is wrinkly. Because *S* is dominant, three of the four possibilities result in a smooth seed.

Mendel's theory was completely ignored until well after his death; it was not consistent with the views of Galton, who saw inheritance as a kind of blending, and it didn't appear to fit with human experience.[21] The traits of most children do not perfectly copy those of one parent or the other, but are usually a mix of the two. However, the gene theory was given a boost in the early twentieth century by experiments on *Drosophila*—those pesky flies that hang around the fruit bowl. The geneticist Thomas Hunt Morgan, at Columbia University, found that the flies inherited traits such as eye colour in the same way that peas inherited their texture.[22] Later, Hermann Müller showed that the flies often suffered severe genetic mutations as a result of X-ray damage to their chromosomes (so named because they absorbed coloured dyes). This implied that the chromosomes were the location of the genetic material—and also, as Müller pointed out, that X-rays were more dangerous than they seemed.

The history of chromosome research is another example of how scientific theories—and basic shape-recognition and counting skills—can be affected by sociological forces. In 1923, Müller's eminent colleague Theophilus Painter announced that, by means of complicated microscopic observations, he had counted the number of human chromosomes, and there were twenty-four pairs.

Other scientists repeated his observations and came up with the same number. Some thirty years later, new methods allowed cells to be placed onto microscope slides, giving scientists a better look at the chromosomes. It soon became obvious that there were in fact only twenty-three of the little fellows. Painter had got it wrong, but his influence was so great that many scientists preferred to believe his count over the actual evidence. Textbooks from the time carried photographs showing twenty-three pairs of chromosomes, and yet the caption would say there were twenty-four.[23]

Mendelian inheritance became widely accepted when it was found that the great majority of traits, including those studied by Galton, are the result of a combination of genetic and environmental factors, which is why they tend to vary over a continuous spectrum. The traits that Mendel had chosen to study were therefore the exception, rather than the rule (as Hermann Bondi wrote of Newton, "his genius *selected* an area where such perfection of solution was possible"). The mathematician R. A. Fisher, who was Galton Professor of Eugenics at University College London, also showed that while individual genes could vary in a discrete fashion at the individual level, the total "gene pool" would vary in the continuous manner assumed by Darwin's theory of evolution.

THE CENTRAL DOGMA OF THE SELFISH GENE

By 1944, the genetic material in the chromosomes had been narrowed down to a "master molecule" by the name of deoxyribonucleic acid, or DNA. The key ingredients were four bases: adenine, cytosine, guanine (first found in guano), and thymine; these were denoted by A, C, G, and T. The quantum physicist Erwin Schrödinger suggested that the letters must spell out some kind of code.[24] In 1953, the American biologist James Watson and the English physicist Francis Crick jointly published the structure of the molecule.[25] Their double-helix model of DNA, confirmed by the X-ray diffraction results of Rosalind Franklin, won them the Nobel Prize; Franklin died at the young age of thirty-seven from ovarian cancer, perhaps caused by mutations in her own cells owing to X-ray exposure.[26] The unravelling of the structure of DNA appeared to solve a number of riddles almost at one stroke. It provided an explanation of how genetic information is stored and inherited; how it replicates when new cells are formed; and how it evolves. We learned how the information is translated into proteins, and potentially, how we can predict our future health.

The double helix consists of two extremely long strings of the four bases. The strings are complementary, so a C on one always bonds to a G on the other and an A always bonds to a T. DNA is not a single molecule but is divided in humans into the twenty-three pairs of chromosomes. One of each pair contains genetic information from the mother, one from the father. Every cell in our bodies therefore contains two versions of the DNA story, with the exception of sperm or egg cells. In these, the genetic information is compressed into a single strand containing only a single copy; when egg and sperm merge at conception, the fertilized egg again has two copies. We therefore inherit a mix of genetic information from each parent, as Mendel described.

When a cell divides to form a new copy, the two strands of DNA are unzipped and a complement to each strand is produced by adding the missing bases. The result is two copies of the original. The copying process is tightly controlled and most mistakes quickly repaired, so each human cell normally has only a few errors. New DNA errors are continuously introduced by external effects such as ultraviolet radiation or carcinogenic chemicals, however; sunlight and cigarettes both go right to the DNA. It's a molecule that is always patching itself up.

Despite this constant grooming, mistakes happen. A base is left out or a new one put in. If such point mutations happen in egg or sperm cells, they can be passed on to the next generation. Sometimes, this can result in serious genetic diseases; however, random mutations are also one of the drivers of evolution, since occasionally they provide a beneficial feature (like these fins on my back).

While DNA seemed to be the "master molecule" of heredity, it had long been known to biologists that the workhorses of the cell, which gave it structure and carried out the metabolic processes that sustain it, were strings of amino acids known as proteins. The problem of how the information in DNA is translated into proteins was cracked in the 1960s by a group of scientists that included Crick and a team at the Pasteur Institute in Paris. They showed that each three "letters" of DNA, such as the string ACT, represents one of the twenty different amino acids. Mendel's factor—or gene—was a segment of DNA, consisting of perhaps thousands of letters, that codes for an entire protein.[27]

DNA ———▶ RNA ———▶ Protein ———▶ Organism

FIGURE 5.3. The central dogma. Genes encoded in DNA are transcribed to form molecules of RNA, which are translated to form proteins, which carry out the work of the organism.

Proteins are not produced directly from DNA. In an intermediate step, a molecule of RNA—a complement to DNA (like a photographic negative), but containing only the information relevant to the particular protein—is transcribed. The RNA is then translated into a string of amino acids, which must be folded into shape to form the actual protein. It is the shape of a protein that in large part determines its behaviour (like how it bonds with other molecules). At the molecular level, form and function are the same. The transcription of RNA—and its translation into proteins—is another tightly controlled process. Proteins that misfold or are flawed in any way are targeted for destruction by the cell—though different proteins may be folded from the same amino acid string.

Francis Crick dubbed the DNA-RNA-protein model the "central dogma" of molecular biology—a Newton's law for life. This highly stable, linear, one-way information flow has since been revealed to be considerably more complicated. Barbara McClintock showed that genes are themselves dynamic and can jump from chromosome to chromosome (or even species to species). The RNAs produced were also found to go through substantial splicing and modification before being translated. DNA is like a first draft that is reshaped by a very proactive editor. Evolution is therefore influenced by non-genetic factors, as well as random mutations.[28] However, the central dogma was extremely successful in explaining the basic mechanics of inheritance—so successful that biologists like Richard Dawkins began to see life forms as little more than glorified delivery systems for their "selfish genes."[29]

DNA, it seemed, was rather like a kind of software program for the expression of genes. If scientists could figure out the operating system, then perhaps they could hack into the program to read the future—and even change it.[30]

THE MASTER MOLECULE

You can easily inspect the DNA molecule in the comfort of your own home.* Take half a cup of split peas (with a nod to Mendel), add a pinch of salt and a cup of ice-cold water, and blend on high for fifteen seconds. This separates the pea cells. Strain the soup and add about two tablespoons of liquid detergent. Mix, then let sit for ten minutes. The detergent breaks open the cell nucleus, which holds the DNA. Pour the liquid into test tubes or similar glass containers until they are about a third full. Now add a pinch of meat tenderizer to each tube and stir very gently. The tenderizer includes enzymatic proteins that break up the proteins clustered around the DNA. (If you don't have meat tenderizer, pineapple juice may work.) Finally, tilt the container and slowly pour rubbing alcohol down one side, as if you were carefully trying to fill a glass of beer, until the volume has doubled. If everything has gone as planned, some white stringy stuff will gently rise to the surface of the alcohol, which you can extract using a chopstick or whatever you have at hand. This is the DNA. It's stringy because the long DNA molecule is partly unwound. A molecule of human DNA, if uncoiled, would be almost two metres long, though of course immensely thin.

If you like, carefully rinse the DNA and transfer it to a fresh glass. Add a shot of vodka, ice, and tonic. This Master Molecule drink will bring any party to "life"—at least a primitive, pea-like form of life.

Based on a recipe from the Genetic Science Learning Center, University of Utah.

Big Science Meets the Master Molecule

In the early 1980s, the U.S. Department of Energy was casting around for so-called big science projects that would complement its high-energy physics programs. In December 1984, at a summit to discuss techniques to detect genetic mutations in descendants of the survivors of Hiroshima and Nagasaki, someone proposed a project to determine the entire sequence of the human genome—the exact order of the As, Cs, Ts, and Gs (which differs slightly from person to person). The sheer length of the molecule meant that it would be impossible to decode in a reasonable time without the invention of new technologies. However, the promise seemed enormous: knowledge of DNA could help (or at least identify) victims of genetic diseases, lead to new kinds of drugs, and perhaps reveal the secrets of life.

The Human Genome Project, as it became known, has been compared to determining the exact address of each person on the planet. If the earth corresponds to the genome of a single cell, then a chromosome is a country, a gene is a town, and the base pairs—those billions of As, Cs, Ts, and Gs—are individual people. The enormous task was made possible by a range of new techniques for determining the sequence of DNA, which worked essentially by dicing it into small chunks, figuring out the sequence for each piece, and splicing it all together again on a computer. Attracted by the emphasis on numerical approaches, as well as the funding possibilities, many physicists, engineers, computer scientists, and mathematicians switched their attention from physical systems to living ones and joined the search for the so-called biological grail. (Of course, according to mechanistic science, there is no fundamental difference between physical and biological systems.) New disciplines were born, including bioinformatics, which uses mathematical and statistical tools to probe strings of DNA for hidden meaning, and systems biology, which analyzes the function of complex biological networks.

Rather than start straight in with the human genome, it was

decided to warm up on some simpler organisms. The bacterium *Haemophilus influenza* had its DNA decoded in 1995. Baker's yeast, *Saccharomyces cerevisiae,* gave up its secrets in 1996.[31] The effort to determine the sequence of human DNA turned into a competition between a private company, Celera, headed by the former professional surfer Craig Venter (who sequenced DNA from his own body, perhaps to find the elusive surfing gene), and a public consortium of universities and research institutes. After a sometimes acrimonious battle—DNA, it seemed, was not just the molecule of identity, but the molecule of ego—the two sides eventually patched up their differences and, to great fanfare, jointly announced their triumph in 2001. The British prime minister, Tony Blair, said that it heralded "a revolution in medical science whose implications far surpass even the discovery of antibiotics."[32]

One of the immediate surprises was that the human genome has only about 30,000 genes. The roundworm has about 19,000 genes, and even yeast has 6,000. Most estimates for humans, perhaps out of sheer pride, had put the number at at least 100,000. How could such a small number of genes produce so much complexity? Actually, it's not the number of genes that count, but the different ways they can be combined and expressed, which is essentially limitless. The variability among humans involves even fewer genes: about 93 percent of those discovered are held in common, so individuals may differ in only a couple of thousand genes. But even this is far more than enough to ensure that no people apart from identical twins have the same genes—just as a phone system with only ten digits is enough to supply everyone on the planet with a unique number.

Another surprise was that about 98 percent of the DNA didn't seem to code for any gene at all.[33] The remainder was termed, somewhat prematurely, "junk DNA." At least some of it has since been shown to play an important role in gene regulation, and perhaps even social attributes.[34] The most embarrassing discovery, though, was that the small amount of DNA that does code is about 98 percent the same as a chimpanzee's. We are more than just descended from apes, as Darwin showed; according to our DNA, we practically *are* apes.

Of course, we run into the problem of metric again. A measurement of the genetic text does not correspond to a measurement of particular outcomes. For example, the following two sentences differ by only a few percent in terms of their letters, but their meaning is entirely different:

> Chimpanzees and human beings are very similar animals.
> Chimpanzees and human beings are not very similar animals.

As in language, details count.[35] There isn't a straightforward, linear relationship between letters of code and the end result. The path between the two is crooked. In genetics, the word "not" corresponds to a gene that turns off transcription of another gene. Genes that activate or repress other genes in this way are ubiquitous.

PREDICTING HEALTH

One of the main goals of the genome project is to prevent diseases before they happen.[36] The cost of sequencing DNA—estimated at $3 billion for the first human genome, or a dollar per base pair—continues to plummet. In the near future, it's expected to reach a level affordable to many people—say, under a hundred dollars. Since DNA is present in all cells, it would only take a sample of blood or a swab from the inside of the mouth to determine someone's genetic sequence. Small differences in the genome may correlate to susceptibility to certain diseases, anticipated longevity, and so on. Dozens of companies, including Celera, are searching for DNA tests that can foretell your future health—a genetic horoscope. As James Watson asserts: "We used to think our future was in the stars. Now we know it is in our genes."[37]

Computer-based biological prediction is a new area that is evolving fast. Most genetic tests performed on a routine basis today are for so-called Mendelian diseases—those caused by a single gene. These can be devastating but are fortunately quite rare. Cystic fibrosis, for example, is associated with a recessive gene; if both parents carry it, the fetus may develop the disease. Detection is complicated by the fact that the relevant gene is extremely long, and the corresponding protein exists in hundreds of slightly mutated forms. Nevertheless, genetic testing of the parents can provide some warning.

Huntington's disease is caused by a mutation that inserts too many repetitions of the sequence CAG (code for the amino acid glutamine) into a particular gene. The result is a deformed protein that slowly accumulates in the brain, leading to progressive loss of movement control and eventually death. The age at which neurological symptoms first appear depends on the number of CAG repeats.

The predictive powers of the scientists seem almost magical: by analyzing a single gene, they can foretell a patient's future symptoms. Unfortunately, the magic does not yet extend to a cure. If you have the defective gene, it appears in every cell in the body, so it cannot easily be corrected (though gene therapies hold potential). This raises a number of ethical issues around genetic testing. If your child had a high likelihood of contracting an incurable disease, would she really want to know? Would she want you to know? Would either of you want an insurance company or an employer to know? And when does genetic testing cross the line into eugenics?

The genes *BRCA1* and *BRCA2* were so named for their connection with a rare form of early-onset breast cancer. About 10 percent of breast cancers are believed to be hereditary, and about half of those may be associated with these two genes, which produce proteins responsible for DNA repair. Certain mutated forms are less efficient at this task, and can lead to breast or ovarian cancer. Genetic screening for the condition is complicated by two factors. First, each gene exists in hundreds of different forms, and it is hard to know which are dangerous, so the aim is usually to find out whether a family member with cancer has passed on a copy of the same gene version to their offspring, and the results must be carefully interpreted by a genetic counselor. Second, these genes are "owned" not by the person being tested, but by a Salt Lake City biotech firm called Myriad, who were the first to locate, sequence, and patent-protect them. Myriad therefore has a temporary monopoly on the genetic tests, which again raises a host of issues, especially for those who can't afford the fees charged (typically thousands of dollars).[38] Similar patents, covering the majority of the human genome, have been taken out by other companies including Celera, Human Genome Sciences, and Incyte in a kind of DNA land rush, with the aim of producing useful and marketable genetic tests.

Genes, of course, do not just cause diseases; in fact, they may even grant immunity. A person with blood type AB (about 4 percent of the population) is virtually immune to cholera, for example. Cholera is not a genetic disease, but it can have a genetic cure (or an economic one—it was eliminated from Europe and North America by improvements to the water supply). The gene that causes sickle-cell anemia, a blood condition that is especially common in people of African descent, also protects against malaria. Mutated forms of the gene *CCR5* appear to foil the virus that causes AIDS.[39]

NATURE, NURTURE, OR NEITHER

If a particular gene is mutated, its protein products may cause Mendelian diseases; however, we cannot conclude from these special cases that every disorder has some unambiguous genetic root. Most diseases that have a hereditary component—including forms of cancer, heart disease, asthma, and diabetes—are not the result of a single mutant gene ticking away like a time bomb. Instead, they come from a combination of genetic and environmental factors. (It is often more appropriate to discuss gene versions rather than mutants. Genes vary, and the fact that your DNA differs from Craig Venter's doesn't mean that it isn't normal, or that you are a mutant.) Predicting the appearance of these diseases is not a harder version of predicting the Mendelian diseases, but a completely different—call it Galtonian—class of problem. Just as global political problems are not usually caused by a handful of people and atmospheric storms are not stirred up by a single butterfly flapping its wings, so most health conditions cannot be traced back to a few rogue base pairs.

Heart disease, for example, is influenced to an extent by a gene known as *APOE*. This produces a protein that plays a role in the complex process of preventing cholesterol and fat from accumulating on the linings of blood vessels. The gene comes in slightly different versions, the most common of which are designated *E2, E3,* and *E4*. A person with two copies of *E4* (i.e., who has inherited one from each parent) stands a relatively high risk of developing early heart disease.[40] What counts is how the many different proteins involved interact as a group, however, so no single gene tells the full story. Current research therefore focuses on finding sets of genes that tend to vary together; but it is not clear that these are the sets responsible for disease.[41] There are also other hard-to-quantify and inter-correlated factors that affect heart disease, such as fitness, obesity, noise pollution,[42] tobacco use, stress, job status,[43] the extent of one's social network,[44] the number of fast-food restaurants in the neighbourhood,[45] mental attitude, and of course, sheer luck.

Another challenge in making predictions based on genetic data is that what counts is not the genes we have but the genes we express as proteins. This changes with time, environmental cues (such as fast-food restaurants), and location in the body. All cells in our body contain the same set of DNA instructions, but a liver cell is not the same as a heart cell, and a skin cell is different from a brain cell. They each interpret the text in a different way, using only what they need. In one particularly striking experiment, scientists transplanted mouse cells into chick embryos, and thus induced some of the chicks' own cells to develop teeth. The strange thing is that while chicks normally don't have teeth, they are thought to have descended from dinosaurs, which did. Evolution had repressed the growth of teeth (it had inserted the word "not"), but the genes for them had remained, like a memory from a past life.[46]

The expression of genes is not static and fixed. It's promoted or repressed in a time-dependent, dynamic fashion by proteins that bind directly onto particular sites on the DNA. A gene may be switched on or off by the action of several proteins working in concert. Similarly, hormones such as those associated with stress exert their effect on the body by controlling gene transcription. After stress hormones are released from the adrenal glands, they enter into cells, combine with special receptor molecules, and bind to the DNA, where they control the expression of certain genes. Such hormones are, of course, particularly susceptible to environmental effects. Scud missiles were mostly ineffective against Israel during the first Gulf War, but the threat of them caused a huge spike in the number of heart attacks. Generally, stress is believed to weaken the immune system and make people more vulnerable to disease.[47] Stress acts at a genetic level, but we wouldn't say it was a genetic disease. Mental states such as depression also affect the expression of certain genes, but this does not mean their cause is rooted in those genes.

Even a developing fetus, safely ensconced inside the womb, is subject to complicated environmental effects, although the environment is now the mother's body. One factor is the prenatal testosterone level, which affects the development of the nervous system and is influenced by the fetus's sex. High testosterone levels loosely correlate with autism, immunosuppression, and aggression. On the bright side, those affected are often good at music and mathematics. Prenatal testosterone also influences expression of the genes that control digit length. In women, the ring and index fingers are usually around the same length, but in men the ring finger of the right hand is usually longer than the index finger.[48] Therefore, finger length is a weak predictor of these traits, though the correlation is so weak that it is meaningless to read anything into it for a particular individual. (It has a certain Pythagorean logic, however: a long straight finger on a male right hand is linked with music, mathematics, and aggression.) Since the prenatal testosterone level is part of the dialogue between mother and developing child, it is impossible to say whether it is nature or nurture.

Practically all traits, in fact, are the result of interplay between nurture and nature. A child might be anxious because of an inherited nervous temperament (nature), or because of exposure to stress-causing experiences (nurture). But that nervous temperament might have been induced not by the parents' genes, but by their anxious behaviour (nurture). Or those stress-causing experiences might be the result of an inherited tendency to seek risk (nature). Or the parents' anxiety might have been caused by the child's behaviour, making the child even more stressed, and so on, in a positive feedback loop, until the men in the white coats show up. Nature and nurture are like two intertwined forces that sometimes work together and sometimes work in opposition. Even the nurturing experience of reading this book will change neuronal pathways in your brain, with perhaps strange and unpredictable effects on your nature. And to nature and nurture, we must also add blind chance—those subtle effects that are essentially random and cannot be controlled.

Chance plays a role in every stage of our lives, not least the moment when we are conceived. Aristotle believed that a child's sex was determined by the temperature of the male's sperm, which he in turn thought was a function of the weather. He therefore advised people who wanted a male heir to conceive during the summer.[49] While the sex of some reptiles does depend on the temperature at which the eggs develop (the preponderance of males born in some species is an early indicator of global warming), a human child's sex is determined by a random process—namely, whether the sperm that fertilizes the egg contains an X chromosome (female) or a Y (male).[50] Similarly, a coin toss may determine whether a particular gene is expressed, a neural pathway laid down, or a cancer developed. And owing to small random differences in their development, even genetically identical twins are not really identical, as their fingerprints, those weather maps of identity, will show.

DATA OR MODEL

As mentioned above, scientific forecasts can be based on either statistical patterns or detailed mathematical models. An inherent drawback of the data-driven approach is that it relies on the Pythagorean idea that the future will resemble the past. Galton showed that tall parents correlate with tall children, but his results applied to a population, rather than to individuals, and were not proof of a causal relationship. Change the population, and an "inherited trait" may weaken or disappear. The same applies to genetic risk factors.

The fact that a gene is statistically associated with a trait in the general population does not mean it will cause that trait in a particular individual. Data also often includes a large amount of random variation or noise, so it is often impossible to tell whether a correlation exists at all. It is therefore not very meaningful to say that a trait is 30 percent inherited, or that a certain gene gives someone a 20 percent chance of developing a disease. To get around this problem, genetic studies often focus on subpopulations with a high degree of genetic homogeneity, such as Icelanders or Ashkenazi Jews, but the results are of most use for those groups.[51]

Statistically based claims that particular genes cause complex diseases usually prove to be oversimplistic. The biologist David S. Moore wrote that "a perfect and complete map of the human genome will not allow us to make accurate predictions about the traits—or diseases—that a given human being will develop."[52] The genes for Parkinson's disease, suicide, and homosexuality have all been discovered and then promptly undiscovered, after further experiments showed that the story was more complicated.[53] Inheritance is obviously important—animal breeders would go out of business otherwise—but individual genes, or limited sets of genes, do not fix our future in stone.[54] Our fascination with finding a molecular basis for complex conditions like mental illness often appears to reflect a cultural desire to explain the world in terms of simple physical causes, which can be harmful if it distracts us from addressing more relevant issues.[55]

The ultimate aim of the Human Genome Project and predictive medicine is more general, and more ambitious, than this. It's to translate the language of DNA and use it to determine the connection between the genotype—the sum of genetic information in the DNA—and the phenotype—the appearance and properties of the final organism. Many top molecular biologists believe that it should be possible to compute the phenotype from the genotype using a mathematical model. Efforts to build such models are under way at a number of institutions. In principle, they would account for all the processes and interactions that make statistical prediction so difficult. They would give an almost unlimited ability to predict future traits based on the laws of cause and effect. They would be the GCMs of the body. One Canadian scientist compared the process with that of weather forecasting: "What they're looking at is a whole system— looking at clouds coming from Alberta and looking at where they will be in a few days. That's basically what we are going to do with the cell . . . . [We'll] be able to predict what's going to happen."[56]

Such models are reminiscent of Laplace, who said that if we knew the initial conditions of all particles and the laws of motion acting on them, we could predict the future state of the system. So is it possible to predict health using DNA? Can we build a model that will forecast traits based on genetic information? Is life computable, or will a baby's future turn out to be more elusive than the clouds? To answer these questions, we must investigate the processes by which information in DNA is expressed in the organism.

THE GENERAL CELL MODEL

To build a predictive, mathematical model of a living system, we must take into account the genotype, the phenotype, and the external surroundings. For DNA to code for proteins, it needs to be part of a body, which in turn is nourished by external resources.[57] A molecule of DNA, left to its own devices, will not spontaneously organize into a life form. In fact, DNA is a highly inert substance, which is why it has taken over from Galton's fingerprints as the police ID method of choice. It can be recovered from tissue, bone, or bodily fluids at crime scenes, and can be used to put people in prison, or get them out, even years after a crime was committed.

To get a sense of the biochemistry involved, we'll follow the lead of the genome project and begin with the simple and unicellular *E. coli.* This tiny bacterium, barely visible with an optical microscope, resides happily in the human gut, among other places, but will also grow in petri dishes supplied with the correct nutrients. If the instructions on how to make the cell are contained in the DNA, then we should in principle be able to use the DNA sequence, the cell contents, and the environment as the initial condition and the basic physical laws as the model. Running the model forward in time, we should see how the cell evolves—just as meteorologists do with the weather.

The cell, the basic unit of life, is a little world in itself. It is surrounded by a layer of lipid molecules. This wall separates the cell from the outside world, but it also contains pores and specialized proteins that can absorb resources, expel waste, and read chemical signals. In *E. coli,* about 15 percent of the cell's mass is protein, 6 percent is RNA, and 1 percent is DNA. As in most cells, water takes up about 70 percent. The cellular material is localized in organized structures. It is no more meaningful to talk about the "concentration" of a protein in water, as is typically done in chemistry, than it is to talk about the concentration of Hawaiians in the Pacific.

One problem with modelling a cell is that it is too large: an *E. coli* cell contains about 2 to 3 million protein molecules. It is also too small. You can go out and measure the pressure of the atmosphere or watch the formation of clouds, but it is extremely hard to observe individual cells without disturbing them (though new nanotechnologies hold the promise of reducing the measuring equipment to a scale suitable for a cell).[58]

To capture the process by which the bacterium's DNA instructions are read, the model must simulate the transcription of DNA to RNA, the subsequent translation of RNA to protein, and the folding of the protein into its final shape. While figure 5.3 (see page 187) looks extremely simple, the reality is more like a complicated industrial process, with many different inputs. Each step involves molecular machines that take up a great deal of the cell's space. The DNA molecule is much longer than the cell, so it's tightly folded and held in place by proteins. For a particular gene to be transcribed, the relevant portion of DNA must be made accessible to specialized polymerase molecules. The transcription process is also controlled by proteins that dock onto binding sites near the gene's DNA. These proteins may either repress or promote transcription, and often act in concert, so that a gene is not transcribed unless several different proteins are correctly bound.

The RNA molecules produced by transcription are comparable in length to the cell and typically number in the thousands. In bacteria like *E. coli,* an RNA molecule begins to be transcribed by ribosomes even before it detaches from the DNA. The ribosomes assemble amino acids into proteins according to the sequence specified in the RNA. Once a protein has been produced, it must be folded into the correct shape to function properly. Since a protein containing a string of perhaps thousands of amino acids can fold into a number of different shapes, specialized chaperone molecules help it form—rather like clowns who twist long rubber balloons into amusing shapes.

The protein then heads off to interact with other proteins, form complexes, and take part in some other aspect of the cell's metabolism (from maintaining the cell's structure or transporting molecules to transmitting information or handling waste). This is when things get complicated, as shown in figure 5.4. Metabolic processes include positive feedback loops that amplify signals, negative feedback loops that control them, cascades of reactions that transmit signals, and so on.

While a bacterial cell is certainly a dynamical system—the word "metabolism" is derived from the Greek *metabole,* for change—its dynamics do not resemble those of the planets, where the normal rules of physics can be applied. They are more like those of a city, with a myriad of players engaged in every kind of activity—industry, energy, consumption, garbage disposal, management, maintenance, communication, security. The system becomes even more complicated in eukaryotes such as yeast. In these the cell is divided into a number of specialized compartments known as organelles. DNA is constrained inside a nucleus, where transcription takes place, and proteins are transported from compartment to compartment inside coated vesicles. They ride the bus to work instead of walking. Multicellular organisms like mammals have completely separate layers of complexity.

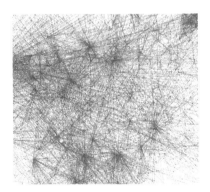

Figure 5.4. Interactions between different proteins in a yeast cell. The nodes represent different types of protein, and the lines between two nodes mean those proteins interact (by joining to form a complex, for example). Systems biologists try to make sense out of this.[59]

Clearly, modelling even a unicellular bacterium such as *E. coli* is an enormous challenge. The cellular machinery lacks what the Pythagoreans referred to as "the universal harmony and consonance of the spheres." It is hard to imagine determining even the correct initial condition for the cell, since this would have to include its entire structure, the configuration of proteins, and so on. The only option is to pass over all these details, make a number of gross simplifications, and hope that the resulting low-resolution model will capture the essence of the underlying dynamics. The model could then be used, for example, to determine how changes to particular genes affect the phenotype—information that might prove useful in the study of human disease.

COMPLICATIONS

The simplest way to build such a model is to treat the cell like a test tube full of chemicals interacting according to the rules that govern large-scale chemical reactions. Because it is not possible to track every single reaction, the model must focus on an isolated subset of the complete system. The result is a large set of differential equations that can be solved on the computer. Such an approach, while straightforward, has a number of drawbacks. The use of differential equations assumes that the chemical concentrations vary in a continuous way. However, RNA molecules are produced from DNA and later degrade in a discrete fashion, so their numbers will vary in a somewhat random manner. Complex organisms use multiple control systems to reduce the effects of this stochastic noise, as it is known (the word "stochastic" comes from the Greek *stokhos,* which was a pointed stake used by archers as a target). Stochastic noise can be captured to an extent by modelling the reactions as discrete, random events between individual molecules, which is more realistic but computationally expensive.[60]

Another modelling difficulty is that measurements of reaction rates and chemical properties in a laboratory setting may be highly misleading, because cellular context matters. If solutions containing two chemicals are mixed together in a test tube, the reaction rate depends on the mixing process and will obey simple statistical laws. In the cell, however, protein traffic is carefully guided and controlled. Therefore, techniques that can be used to model chemical reactions for millions of molecules in a solution are simply not appropriate inside a small cell, where interactions are controlled by local effects. Finally, a model that focuses on only a subset of reactions will have to omit a large part of the system, and perhaps miss important interactions.

Modelling of biological systems is an area of intense research activity, and new algorithms and techniques are constantly being developed. However, as modellers add more detail—say, by simulating the fine structure of the cell—they run into problems trying to verify it properly against experimental data. In biological models, there always seem to be far more parameters than there are quantities that can be measured, and as more detail is added, the number of parameters explodes. It is therefore impossible to deduce their correct values from experiments. The models suffer from the same problem as the Greek Circle Model did: they are too flexible.[61] As Will Keepin put it, modellers can pull the levers and make the model do whatever they want. This is common to all the systems discussed in this book. It is the signature of uncomputability.

LOCAL OR GLOBAL

The more closely we study cellular dynamics, the clearer it becomes that they depend on a multitude of local effects, acting on individual molecules, that cannot be neatly captured by equations. The initial condition is given by the DNA and the starting distribution of proteins and other molecules, and the rules are the laws of chemistry and physics. Yet even perfect knowledge of both these things would not allow us to predict the system's state after a certain length of time. For that, we'd need to run the system itself. The cell's characteristics, or traits, are analogous to emergent properties of complex systems, which elude computation (see figure 5.5 on page 207). It is easy to devise models that appear to fit past data, but this does not mean the models can predict the future.

This problem is compounded in multicellular organisms, like humans. We contain about 100 trillion cells, in a hierarchy where each level depends on the level below and above it. The body is made up of organs, which in turn are made up of separate components, which are made up of cells, which are made up of smaller organelles, and so on, down to the level of molecules. Since there is a constant flow of information between these levels, it is not possible to draw an arbitrary line at some level of complexity. Like clouds or turbulent flow, living beings exhibit structure over a wide range of scales.

It may seem surprising that DNA cannot be read like a book: after all, it is just a string of information, a long text. If we can arrange the cloning of a sheep from a string of DNA, surely we can build a computer model that will predict a phenotype from a DNA sequence. Even the miraculous development of a human being—through the fixed stages of conception, embryo, fetus, and birth—seems to follow a regular, repeatable path, as if carrying out a set of detailed instructions. But the development of multicellular organisms owes less to dictates from a "master molecule" than it does to many small, local decisions. Specific proteins, known as adhesion molecules, cause cells to cluster together or to slide along one another, resulting in the surfaces and folds that define tissues and organs. As the biologist Richard Lewontin put it, "At every stage it is the local interactions of cells and tissues that determine the further movement, division, and differentiation of cells in the locality, which lead to yet further local interactions, and so on to adulthood."[62] The fact that this process appears to be machine-like and, to a degree, reproducible does not mean that it is predictable. If the Game of Life is started twice from the same initial conditions, it will evolve in exactly the same way, but that is of no help at all in guessing its future.

LOCAL RULES	GLOBAL BEHAVIOUR
H_2O	Clouds
DNA	Complex traits

FIGURE 5.5. You can't get there from here. Water molecules and DNA may obey simple, locally applied physical laws, but that does not mean we can compute the properties of clouds or organisms.

NEGATIVE AND POSITIVE

The lack of a direct, computable link between genotype and phenotype means we cannot translate DNA from the bottom up. To find out what a baby will look like, or how healthy it will be, we will still need to wait around and see—it won't suffice to plug the baby's DNA sequence into a computer. It is still possible to build mathematical models using a top-down engineering approach, and these models are useful in understanding the processes behind human health. However, their accuracy is limited by another factor, which has to do with the nature of living systems. Again, the example I give is unicellular, but the principles apply to more complex organisms—indeed, to all the systems studied in this book.

One genetic network that has come under much scrutiny from biologists in recent decades is the metabolism in baker's yeast.

When a yeast cell's food supply switches from its preferred carbon source of glucose to a slightly different sugar, known as galactose, the yeast turns on a particular set of genes, which then produce proteins that digest the galactose. As an experimental organism, yeast has a number of favourable qualities. This microscopic fungus was probably the first organism to be domesticated by man. Without it, we would have neither bread nor beer. It is easy to grow, reproduces quickly, and shares surprisingly many genes with humans. Indeed, about a quarter of the 5,000 or so genetic diseases that afflict humans have some kind of analog in yeast—including galactosemia. A child with this condition cannot digest the galactose in milk, so must avoid it or risk death from the buildup of toxic metabolic by-products. The condition can be treated by a change of diet.

A common feature of genetic networks is auto-regulation, in which proteins regulate their own transcription. There are two types, positive and negative, which correspond to positive and negative feedback. In electronic circuits, feedback loops are used to amplify or damp electrical signals, and they play a similar role in biological systems. Positive feedback, for example, is used to provide a rapid response to a signal, but it has to be carefully controlled. If you cut your finger, a cascade of reactions incorporates positive feedback to create a clot and stop the bleeding. If this system goes wrong, it can cause thrombosis.

The galactose network incorporates both types of feedback loops, but in opposition. When galactose is detected, the yeast cell uses positive feedback to accelerate the production of the proteins required to metabolize galactose. But at the same time, a protein responsible for repressing transcription of the same proteins is also upregulated. This acts as a partial brake on the system—negative feedback. Both the brake and the accelerator are applied at the same time, as if the organism is ramping up its internal tension. This type of antagonistic action is ubiquitous in biological systems; the level of glucose in blood, for example, is controlled by the hormones insulin and glucagon, which pull in opposing directions, like the two hands of an archer.[63] If the hormones get out of balance, it can lead to diabetes. Our appetite is suppressed by one hormone, obestatin, and boosted by a second, grelin, which are both produced from the same gene (the RNA is edited differently).[64]

At the Institute for Systems Biology in Seattle, our computational group—a biologist, an ex-astrophysicist, a mathematician, and an ex-electrical engineer (the group leader, Hamid Bolouri)—produced a dynamical model of the galactose network.[65] The model included fifty-five types of reactions between a range of chemical species, including DNA, RNA, and proteins; stochastic simulations, which tracked interactions between millions of individual molecules, took days to run on a cluster of computers.[66] Complex processes—such as the regulation of DNA transcription, the interior dynamics of the cell, and so on—were represented by simple parameterizations. Such models are useful in the scientific process: like an architectural model of a complicated structure, they help the scientist see how things fit together and test whether her assumptions are reasonable.

The aim of our model was to view the galactose network as a piece of engineering and "predict" the role of various features (many of which are also employed in our own bodies). It appeared, for example, that the intricate combination of feedback loops made the yeast less erratic in its response to galactose. This hypothesis was backed up by a laboratory experiment that compared the metabolic responses of two types of yeast: the regular baker's yeast, and a mutant version whose feedback loops had been disabled using the techniques of genetic engineering.[67] For both strains, there was a spread owing to random stochasticity, or noise, with some individual cells reacting more and others less; however, the simpler mutant strain showed a broader, more erratic response, while the regular yeast was more controlled and robust (see figure 5.6 below). If these were archery scores, we would say that the regular strain was a steadier shot.

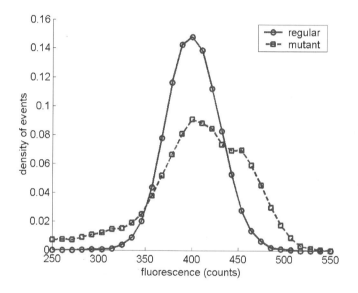

FIGURE 5.6. Histograms comparing the reaction of regular (wild) and mutant yeast to galactose. The horizontal scale indicates the metabolic response six hours after the yeast was exposed to galactose. The mutant strain has a broad distribution, indicating that some cells underreact and others overreact, while the regular type is less erratic. If the distance from the centre of the horizontal scale denoted the distance from a target, the regular strain would be more consistently accurate.

GLOW-IN-THE-DARK YEAST

Its robustness makes yeast hard to model but very convenient for experimentalists and bakers, who can starve it, store it in the fridge, and subject it to all kinds of abuse before miraculously reviving it by putting it in a more welcoming environment (like a petri dish full of nutrients or a bread dough). Biotechnologists go so far as to inject foreign DNA into the yeast genome, to alter certain genes or show when they are being expressed.

Such experiments show the Promethean nature of modern biology. Stretches of DNA can be bought ready-made from biotech companies and inserted into the cell nucleus. One method uses a jolt of electricity, Frankenstein-style, to shock the yeast into opening its pores. In some cases, the DNA is then taken up by the cell's own mechanisms and incorporated into the genome. These cells are then bred in petri dishes to create a modified yeast strain.

One way to tell whether a gene is being expressed is to insert the gene for green fluorescent protein (GFP), taken from a type of jellyfish, and ensure that its expression is regulated like that of the gene in question. To produce figure 5.6, the GFP gene was turned on at the same time as one of the genes in the galactose network. When cells eat and metabolize the galactose, they glow green under ultraviolet light. It is also possible to knock out particular genes so they are not expressed at all, or to make them express at specified levels. Such experiments can be used to test predictions about the function of biological systems.

With simple organisms like yeast, it is possible to do controlled experiments to test hypotheses. However, we must distinguish between this type of general prediction and more accurate "point predictions" for individual cells or groups of cells. The positive and negative feedbacks of the galactose network are in dynamic, non-linear balance, and any system that places powerful forces in opposition to each other is inherently hard to model, because an error modelling one of the forces would throw the system out of balance. Small changes in the model produce large effects. The problem is not sensitivity to initial condition (the galactose model is not chaotic), but sensitivity to small changes in parameterization. This property is illustrated for a simple mathematical system in Appendix III (see figure A.5 on page 360).

Because the galactose model parameters could not be precisely known, they had to be selected and balanced against one another to give plausible behaviour in the first place. Changing them randomly, by even a small amount, would throw the model off course and give unrealistic protein concentrations. If a skeptic had insisted that we quantify the uncertainty in the model, by running an ensemble of forecasts with random changes to the parameters, for example, this would have been a misuse of the model, because the very feature that made it robust—the balance of positive and negative feedback loops—also forced us to carefully adjust (i.e., fix) the parameters.[68] The model could be used to propose testable hypotheses about the function of the network, which was its aim, but it could not be taken literally as an exact simulation of the real system. As we'll see in the third part of the book, similar issues are encountered when modelling the positive and negative feedbacks that regulate the earth's climate.

While the complexity of the galactose pathway is inconvenient for mathematical modellers, it seems entirely reasonable when we consider that yeast cells have evolved over billions of years to survive in a harsh, unpredictable environment. Complex organisms like yeast depend on homeostasis, which keeps their systems in balance.[69] Their rate of metabolism, for example, cannot be allowed to vary over too wide a range. To maintain such stability and direction in a shifting, unpredictable environment—to be a steady shot—they must develop sophisticated systems to control noise and must be able to respond in a flexible way to external perturbations. Positive feedbacks (which allow rapid reaction) must be balanced by negative feedbacks (which pull back towards equilibrium). From the outside, while at rest, it may appear that there is little going on, but in fact this peace is the result of a truce between powerful internal forces. The situation was perhaps best described by Heraclitus (who was a critic of the Pythagoreans and their good–evil polarity): "what is at variance comes to terms with itself—a harmony of opposite tensions, as in the bow or the lyre."[70] This tug of war leads in the organism to a balance between creativity and stability, and in models to sensitivity to parameterization. Because complex molecular processes are computationally irreducible, any model parameterization will include errors, which this sensitivity will magnify.

The sense of a dynamic balance also gives an understanding of why individual genes, or even groups of genes, are only partially useful for predicting an organism's traits. What counts is the behaviour of the whole organism. If any one part is out of balance, then the organism may either adjust itself internally or modify its behaviour to compensate.

One could therefore say that complex organisms are unpredictable because they have evolved to be that way. It's in their nature. As Pasteur Institute biologist Antoine Danchin writes, a fascinating quality of life is that "even if we do not deny its deterministic character, what we can know about it *does not enable us to predict its future.* Life is simply the one material process that has discovered that the only way to deal with an unpredictable future is to be able to produce the unexpected itself" (his italics).[71]

This does not imply that all organisms evolve towards greater sophistication—the preponderance of life on this planet is unicellular. Still, simple organisms like bacteria can retain the element of surprise, but only by constantly shifting and changing shape. Nowhere is the race for prediction seen more clearly than in the struggle between the human immune system and its microscopic invaders. We return to this subject in Chapter 7.

WATCHING FOR STORMS

Since the development of most human traits from the genotype cannot be predicted either from first principles or from observed trends and correlations, it follows that genetic information can provide only a hazy and incomplete picture of human health. Human traits are like the emergent properties of Class IV systems, which defy computation.[72] Still, advances in biology will be immensely useful for making certain types of very exact predictions. Just as a new communication technology (the telegraph) led to the development of synoptic weather forecasting, new medical technologies will advance the detection and diagnosis of diseases such as cancer.

Cancers occur when mutations in a cell's DNA cause it to grow and divide in an unregulated fashion. The particular type depends on the cell (e.g., a skin cell or a blood cell) and on the exact mutation. Microarray technology, which uses nucleic acids to fish for a range of RNA molecules or DNA segments in a prepared sample, can automatically determine what genes are being expressed by a cancer.[73] Its DNA can be identified just like a criminal's. The newer field of proteomics aims to do the same thing for proteins. In some cases, this knowledge will help determine who will benefit from particular therapies. This is valuable to doctors and patients, as well as drug companies (especially since the cost of developing a new drug can be hundreds of millions of dollars).[74] Cancer is nature's reductionist, and increasingly it can be confronted and cured using reductionist science. It is the medical equivalent of looking over the horizon for storms—and it has the same promise to save lives.

It is already possible to purchase a genetic risk-assessment test that predicts our susceptibility to a wide range of diseases. These predictions are based on statistical correlations of certain gene forms with conditions like heart disease or cancer; they may also help doctors tailor treatments or diets to the genetic profile of a particular patient.[75] Such advice should be interpreted with care. Broad statistics don't necessarily apply to an individual with a unique history, environment, and mix of genes. As the late biologist and science writer Stephen Jay Gould said, while himself fighting cancer, "If both genes and culture interact—of *course* they do—you can't then say it's 20 percent genes and 80 percent environment. . . . The emergent property is the emergent property and that's all you can ever say about it."[76] Also, risk assessments become relevant (rather than something to obsess over) only if—like the dangers of smoking or not wearing seatbelts—they are large enough to make it worth modifying our behaviour.

Personal genetic information will have to be closely protected—like any other medical or identifying data—but the current anxiety about the potential misuse of genetic testing seems to be based on an exaggerated idea of its effectiveness. As Columbia University's Joseph Terwilliger has said, "In many ways, scientists overhyped the information in the genome, or at least what we know about it, to the point where now people are getting unnecessarily nervous about societal implications."[77] Genes are not destiny, and their omens and portents are usually vague and sometimes deceiving. Their role is similar to that which Johannes Kepler assigned to the stars: "In what manner does the countenance of the sky at the moment of a man's birth determine his character? It acts on the person during his life in the manner of the loops which a peasant ties at random around the pumpkins in his field: they do not cause the pumpkin to grow, but they determine its shape. The same applies to the sky: it does not endow man with his habits, history, happiness, children, riches or a wife, but it moulds his condition."[78] Like astrology and other forms of prediction, genetic testing and counselling will undoubtedly make extremely good business. A 2003 article in *Nature Genetics* observed that "the predictive power and mystique associated with genetics, consumers' desire to take control of their health and be proactive, and the ease of advertising and ordering tests on the Internet combine to create a powerful incentive for companies to continue developing and promoting genomic profiling regardless of whether the tests have been validated and proven useful."[79]

Most diseases that have a hereditary component are influenced in their development by a broad range of genetic and environmental factors (two of the most important are hunger and poverty—health often depends more on economics than on genetics). These interact and develop in complex, unpredictable, and often antagonistic ways. We shouldn't anticipate the arrival of a magical machine that will fly into someone's future and predict whether he will drop dead of a heart attack in seventeen years. Traits such as personality are even harder to pin down. My newborn child may have a quarter of my father's genes, but I have no idea whether she will have 25 percent of his sense of humour, or even what that would mean. Uncertainty, it seems, is in our blood.

This may seem disappointing. As Lewontin wrote, "Saying that our lives are the consequences of a complex and variable interaction between internal and external causes does not concentrate the mind nearly so well as a simplistic claim; nor does it promise anything in the way of relief for individual and social miseries. It takes a certain moral courage to accept the message of scientific ignorance and all that it implies."[80] As we will see, uncomputability is a stern and unforgiving companion. On the other hand, perhaps it is no bad thing if nature maintains a little mystery—after all, past attempts at genetic prediction have a long, dark shadow side.[81]

Like weather forecasting, the prediction of complex traits or diseases appears to defy our computational ability: there is no Apollo's arrow to fly into an organism's future. We cannot model the trajectory of a life the way we can the arc of a planet across the sky. The issues are perhaps best summarized by John Maynard Keynes, who wrote in 1933, "We are faced at every turn with the problems of Organic Unity, of Discreteness, of Discontinuity—the whole is not equal to the sum of the parts, comparisons of quantity fail us, small changes produce large effects, and the assumptions of a uniform and homogeneous continuum are not satisfied."[82] But Keynes, an economist, wasn't talking about genetics. He was talking about the Great Depression. In the next chapter, we examine how mathematicians and scientists have modelled another great biological system, the economy.

6 ▶ BULLS AND BEARS
PREDICTING OUR ECONOMY

Rules by their nature are simple. Our problem is not the
complexity of our models but the far greater complexity of a
world economy whose underlying linkages appear to be in a
continual state of flux.

    —Alan Greenspan, chairman of the U.S. Federal Reserve
      (1987–2006)

To me our knowledge of the way things work, in society or in
nature, comes trailing clouds of vagueness.

    —Kenneth Arrow, Nobel laureate in economics

ANATOMY OF A STORM

In 1720, an entrepreneur set up a company in England for the mysterious purpose of "carrying on an undertaking of great advantage,
but nobody to know what it is." Five thousand shares of £100 would
be issued. The public could get in at the ground floor by making a
deposit of £2 per share. In a month's time, the technical details of
the project would be filled in and a call made for the remaining £98.
The promised payoff was generous: £100 per year, per share.

    The next morning, the man opened his office in London's

financial district to admit the deluge of investors who were waiting outside. At the end of the day, he counted his takings, saw he had sold a thousand shares, and promptly left for Europe.[1]

In the same year, other company prospectuses offered even less investor value. One tried to raise a million pounds to fund the development of a "wheel of perpetual motion"; others promoted techniques for "extracting silver from lead," and for turning "quick-silver into a malleable fine metal." But the one that caused the most damage, and created the almost hysterical environment for these "bubble" investments, as they came to be known, was a much larger project that was sanctioned by the highest levels of government, all the way up to King George I.

The South Sea Company was established in 1711 by Robert Harley, Earl of Oxford, to help fund the national debt. In return, it was granted a permanent monopoly on trade with Mexico and South America. Everyone knew these places harboured inexhaustible supplies of gold, and investors were easily found. The actual trading was slow to get started—the king of Spain restricted traffic to only one ship per year, and the first didn't set sail until 1717—but the less that was delivered, the more enormous seemed the potential.

None of the company directors had experience with South Seas trade. Safely ensconced in their London office, they managed to arrange some slave-trade voyages, but these weren't particularly profitable. Preferring to concentrate on financial schemes, and fuelled by the public's love of gold, they made a bid to triple their share of the national debt to almost the full amount, at better terms than those offered by the Bank of England. Robert Walpole from the bank protested against the scheme, telling the House of Commons that it would "decoy the unwary to their ruin, by making them part with the earnings of their labour for a prospect of imaginary wealth."

His Cassandra-like warnings went unheeded. Eased along by a number of large bribes, in the form of offerings of stock to politicians and influential people like King George's mistresses, the company's proposal was accepted by the government. Rumours of increasing Latin American trade spread like wildfire through the coffeehouses in Cornhill and Lombard streets, and in 1720 the stock rose yeastily, from £175 in February to over £1,000 by June. By then, scores of other businesses had been set up to cash in on the growing public interest in stock-market investment, which seemed to be making so many rich. As Charles Mackay, in his book *Extraordinary Popular Delusions and the Madness of Crowds,* commented: "The public mind was in a state of unwholesome fermentation. Men were no longer satisfied with the slow but sure profits of cautious industry."

When the stock price crossed the four-figure barrier, the company directors, either spooked or sated, began to sell their shares. The price stabilized, then wobbled, then started to sink, as investors began to suspect they had been the victims of a giant scam. By the end of September, the price had collapsed to £135. As Jonathan Swift wrote:

> Subscribers here by thousands float
> And jostle one another down
> Each paddling in his leaky boat
> And here they fish for gold, and drown.[2]

Parliament was recalled to discuss the crisis. The bishop of Rochester called the scheme a "pestilence," while Lord Molesworth suggested that the perpetrators be tied in sacks and thrown into the Thames. Robert Walpole was more contained, arguing that there would be time later to punish those responsible. "If the city of London were on fire, all wise men would aid in extinguishing the flames, and preventing the spread of the conflagration before they inquired after the incendiaries," he remarked.

As they argued, the company treasurer Robert Knight put on a disguise, boarded a specially chartered boat, and slipped across the Channel to France. The former chancellor of the exchequer, John Aislabie, who was the company's main advocate in the government, stayed to face the music and was escorted, Martha Stewart–fashion, to the Tower of London.

As a response to the crisis, which had caused a record number of bankruptcies at every level of society, Parliament passed the Bubble Act in 1721. It forbade the founding of joint-stock companies without a royal charter—but it didn't manage to ban bubbles, or the occasional "irrational exuberance" of investors (as Alan Greenspan later described it). When the NASDAQ soared to new heights at the turn of the millennium, how many of its listed companies were engaged in an "undertaking of great advantage, but nobody to know what it is"? And could a trained scientist have predicted such rises and falls? Isaac Newton, who lost a large part of his fortune in the South Sea bubble, didn't think so. As he said in 1721, "I can calculate the motions of heavenly bodies, but not the madness of people."

Nonetheless, the gleaming skyscrapers of financial centres like London, New York, and Tokyo are full of professional prognosticators who make good money forecasting the future state of the economy. Is it therefore possible to build a dynamical model of the economy—a kind of global capital model—which is capable of forecasting economic storms? Given enough data and a large enough computer, can we predict the circulation of money just as we predict the orbit of Mars?

MAKING DOUGH

As Vilhem Bjerknes pointed out, the accuracy of a dynamical model depends on two things: the initial condition and the model itself. To know where the economy is going, we must first know its current state. In 1662, a London draper named John Graunt tried to do for his city what Tycho Brahe and other astronomers had done for the heavens: determine its population. His work *Natural and Political Observations Made upon the Bills of Mortality* compiled lists of births and deaths in London between 1604 and 1661. Many of the deaths were attributed either to lung disease, which Graunt associated with pollution from the burning of coal, or to outbreaks of the plague, like the one that forced the young Newton to leave Cambridge for the countryside.[3]

Graunt's book can be seen as the beginning of the fields of sampling and demographics. The aim of sampling is to obtain estimates using a limited amount of information. Graunt used birth records to infer the number of women of child-bearing age. He then extrapolated to the total population, obtaining an estimate of 384,000 souls.

All measurements of the physical world contain a degree of uncertainty and error. To determine the average rainfall in a particular region, for example, we cannot count all the water that falls; instead, we'll make use of a few randomly located rain-collection devices, each of which will give us a different reading. The average of these should be a good indicator of the region as a whole, provided the rainfall is reasonably uniform. When Tycho Brahe tried to measure the location of a distant star, this too was subject to error because of atmospheric distortions and because his measurements were done by eye. His sextant would give one answer one day, a slightly different answer the next. The "true" answer in such a situation is never known, so Tycho and other astronomers would take the average over different measurements. The sampling was done not over different stars, but over different measurement events.

Similarly, when pollsters want to estimate the average yearly income for a certain area, they ask a relatively small sample of people. So long as those selected are representative of the population as a whole, their responses can be used to make a statistical prediction of the likely average.

Given that all measurements are subject to uncertainty and error, how can we ever be sure that the answer we obtain is good enough? This question was addressed by mathematicians such as Jakob Bernoulli. Part of the famously talented Bernoulli clan, which was later studied by Galton as an example of inherited eminence, he imagined a jar containing a large number of white and black pebbles in a certain proportion to one another. We pull out one pebble, and it is black. The next is white. Then a black, and another black. Then three whites in a row. How many pebbles do we need to examine to make a good estimate of the true proportions? The answer was provided by Bernoulli's law of large numbers, which showed that as more pebbles are sampled, their ratio will converge to the correct solution. In other words, sampling works for pebbles, so long as the sample is large enough.

Bernoulli believed that this result could be generalized beyond pebbles. He wrote, "If, instead of the jar, for instance, we take the atmosphere or the human body, which conceal within themselves a multitude of the most varied processes or diseases, just as the jar conceals the pebbles, then for these also we shall be able to determine by observation how much more frequently one event will occur than another."[4] Until then, the laws of probability had been limited to games of chance, where the odds of holding a face card or rolling two sixes, could be computed exactly. Using the techniques of sampling, it seemed, scientists could make probabilistic estimates of anything they wanted.

Sampling methods were put on a still firmer basis by de Moivre's discovery of the bell curve, or normal distribution. His 1718 work, *The Doctrine of Chances,* which was dedicated to his London friend Isaac Newton, showed how, under certain conditions, a sample of random measurements will fall into the bell-shaped distribution, which peaks at the average value. As mentioned in the previous chapter, one of the most enthusiastic supporters of the normal distribution was Francis Galton, who described it as "the supreme law of Unreason" because events that appear random turn out to be governed by a simple mathematical rule.[5] The mean (or average) and standard deviation of the curve can be used to determine the margin of error of a measurement or the expected range of a quantity based on a sample. Because mortality statistics tend to cluster according to a normal distribution, it was soon also used by insurance companies to determine life expectancies, and therefore to price annuities.

Galton's work on inheritance drew on the research of the Belgian scientist Lambert Quetelet, who in his 1835 *Treatise on Man and the Development of his Faculties,* turned normal into a kind of character: *l'homme moyen,* or the average man. He claimed that "the greater the number of people observed, the more do peculiarities, whether physical or moral, become effaced, and allow the general facts to predominate, by which society exists and is preserved."[6] Perhaps this inspired Galton to make his composite photographs of convicts' faces. It also seemed to put the social sciences on a footing similar to that of the physical sciences, which had made great strides by realizing that it is not necessary to model each particle in detail. The temperature of a gas is a function of the average motion of its individual molecules, so in a way it's a measure of the "average molecule." Perhaps a crowd of people could be similarly described by one average person, which would certainly make analysis easier.

Of course, while molecules of air are identical and don't interact except by colliding, the same is not true of people. As the economist André Orléan observed, our "beliefs, interpretations and justifications evolve and transform themselves continuously."[7] If someone knocks on your door, says she is doing a survey, and asks how much money you earn in a year, you may tell her to go away, participate but lie because you are concerned that she is a tax officer in disguise, or say what you think is the truth but be wrong. If she asks how you feel about the economy, the answer may reveal your own strongly held opinions, but equally it may reflect a discussion you had the previous night or something you saw on a recent TV show. It will also depend on the exact way the question is asked. Just as in physics there is an uncertainty principle that states that the presence of the observer affects the outcome of an experiment, there is a corresponding demographic principle that says any answer is subjective and is affected by the questioner, and even by the language used. This is why politicians spend so much time arguing over how questions should be worded in referendums. Even exit polls are prone to error, as we saw in the 2004 U.S. election, when exit polls at first had John Kerry winning over George Bush.

In measuring the economy, about the only thing that can be counted on is money. An individual can count how much he earns; a company's accountants can count how much it has produced; and a government's accountants can count how much it has taxed. (Most societies undervalue things like trees because it is easy to calculate how much they're worth when cut into pieces but much harder when they're left intact. And trees don't have accountants. This bias may change, to a degree, as economists try to invent costs for such "services."[8]) Even counting money, however, is not straightforward. Governments constantly revise important data, such as unemployment figures and gross national product, and creative accounting techniques, like those used to value Internet companies in the 1990s, distort company accounts.

WHAT'S IT WORTH?

Demographics and accounting can give some insight into the current state of the economy, but to know where it is headed, we need to understand the dynamics of society, and in particular of money. Are there simple laws that underpin economics, like an analogue of Newton's laws of motion for the movement of capital?

Such laws would obviously depend on the idea of value. Just as air flows from areas of high pressure to areas of low pressure, money flows through the economy, seeking out investments that are undervalued. The English philosopher Jeremy Bentham associated value with an object's utility—the property that brings benefit to the owner.[9] His follower, the economist William Stanley Jevons, noted that the utility of an agricultural commodity such as wheat depends on the amount available, which in turn is closely linked to the weather. If a harvest is ruined by drought, then bakers and others compete for the scarce resource, driving up the price. Jevons, who also produced meteorological works (including the first scientific study of the climate of Australia), believed that the weather was affected by sunspots. He therefore developed a model of the boom/bust business cycle based on the sunspot cycle. His hope was that economics would become "a science as exact as many of the physical sciences; as exact, for instance, as meteorology is likely to be for a very long time to come."[10]

The desire to maximize utility is a kind of force that drives the economy. Again, though, there's an important difference between utility and a physical property such as mass: the former is a dynamic quantity that for each person depends on her subjective expectations for the future. Financial transactions are based not just on present value but also on future value. In a market economy, where prices are not set by the state, future value is subject to a number of factors.

First, the value of an asset depends on its prospects for future growth. You don't spend the asking price on a house if you suspect that in five years' time its value will be diminished. Similarly, assets such as stocks can be redeemed only if the seller finds a buyer, which can be tricky if the company's market evaporates overnight.

An asset's valuation must also take into account its risk, which is related to its tendency to fluctuate. Suppose that instead of yeasts, figure 5.6 (see page 210) showed the historical returns from two assets, known as Regular and Mutant, over forty years. Since most people like to avoid unnecessary risk, if only so they can sleep at night, an asset that fluctuates greatly in price, like Mutant, is worth less than one that's relatively stable. The volatility, usually denoted by σ, can be calculated from the standard deviation of the price fluctuations—assuming, of course, that these follow a normal distribution and the volatility does not change with time.

Finally, the value of money will also change, because of inflation and interest rates. If a stock pays a dividend of one dollar in a year's time, and if inflation is zero and interest rates are 3 percent, then the "present value" of that dividend is ninety-seven cents (that's how much you'd have to invest to receive a dollar in a year's time). In economics, time really is money.

Calculating the present value of an asset is clearly a challenge. In the case of a stock, you have to estimate the company's rate of growth, its volatility, any dividends, and the interest-rate environment—not just for now but into the future. Since none of these can be known by an investor who is less than clairvoyant, it means that the present value is at best a well-educated guess. Bonds at least pay you back, but only at some future date, by which time the value of money has changed. Even fixed interest-cash deposits are subject to the effects of inflation.

Assets that have limited functional use and do not earn interest, such as gold or diamonds, are considered valuable in part because of their beauty but mostly because of their scarcity. Galileo wrote, in *Dialogue Concerning the Two Chief World Systems,* "What greater stupidity can be imagined than that of calling jewels, silver, and gold 'precious,' and earth and soil 'base'? People who do this ought to remember that if there were as great a scarcity of soil as of jewels or precious metals, there would not be a prince who would not spend a bushel of diamonds and rubies and a cartload of gold just to have enough earth to plant a jasmine in a little pot, or to sow an orange seed and watch it sprout, grow, and produce its handsome leaves, its fragrant flowers, and fine fruit."[11] Adam Smith echoed him a century and a half later. "Nothing is more useful than water: but it will purchase scarce any thing; scarce any thing can be had in exchange for it," he wrote. "A diamond, on the contrary, has scarce any value in use; but a very great quantity of other goods may frequently be had in exchange for it."[12] Since the perceived scarcity depends on demand, it too is subject to the whims of the market.

Value is therefore not a solid, intrinsic property, but is a fluid quality that changes with circumstances. The value of a bar of gold is determined not by its weight but by what the gold market will bear. Value in the end is decided by people, in a social process that depends on complex relationships in the marketplace. It is subjective rather than objective, moving rather than fixed. Indeed, we often seem more sensitive to changes in price than to the price itself (just watch what happens whenever the cost of gas spikes).

Because of this variability, it would seem that the economy could never reach equilibrium. Nonetheless, economists in the late nineteenth century reasoned that if the market were somehow to settle on a fixed price for each asset, which everyone agreed reflected its underlying "true" worth, then the future expectations of investors would align perfectly with the present. Furthermore, any small perturbation would be damped out by the negative feedback of Adam Smith's invisible hand: the self-interest of "the butcher, the brewer, or the baker." If the price of wheat was too high, then more producers would enter the market, driving the price back down. Fluctuations in prices would die out. Just as a molecule of gas has a known mass, every asset or object would have a fixed intrinsic value.

Of course, there will always be a constant flow of external shocks, new pieces of information that impact prices. In 2004, for example, North American bakeries had to deal with record cocoa prices caused by violence in the Ivory Coast; a rise in the cost per kilo for vanilla, owing in part to cyclones in Madagascar; high sugar prices caused by damage to crops in the Caribbean and the U.S.; expensive eggs because of avian flu; record oil prices, which affect transportation costs; and so on.[13] All of these factors would affect a bakery's bottom line, so the market would adjust its predictions about its performance. (Businesses often insulate themselves against such fluctuations by purchasing futures contracts, which allow them to obtain resources in the future at a fixed price.)

These ideas were the foundations for the theory of competitive, or general, equilibrium. It assumed that individual players in the market have fixed preferences or tastes, act rationally to maximize their utility, can calculate utility correctly by looking into the future, and are highly competitive (so that negative feedback mechanisms correct any small perturbations to prices and drive them back into equilibrium). These assumptions meant that the economy could be modelled and predicted as if it were a complicated machine.

PREDICTING THE PREDICTORS

The equilibrium theory saw the *homme moyen* as a stable, tranquil, emotionally dead person, a mere cog in the machine who would be utterly predictable if it weren't for the rest of the world, with its constant stream of random and disturbing news. Every time a news flash arrives, the *homme moyen* responds by fiddling the control knobs on his portfolio. He can always account for his actions later with a cause-and-effect explanation. Louis Bachelier, however, took the idea of randomness a step further. A doctoral student of Henri Poincaré, the discoverer of chaos, Bachelier chose as his thesis subject the chaos that took place at the Paris Exchange, or Bourse, a building modelled after a Greek temple. In his 1900 dissertation, he argued that new information is unpredictable—which is why we call it news—and so is the reaction of investors to that information.[14]

Most information, after all, has a somewhat ambiguous effect on the market. If the U.S. dollar falls in value, this has one impact on oil producers, another on the tourism industry, another on bakers, and so on, so the net effect in an interconnected world is hard to know to complete accuracy. The price of an asset corresponds to a balance struck in a battle between two opposing, almost animalistic, forces: buyers and sellers, bulls and bears. The reaction of different investors to news will depend on their own subjective interpretation of events, and "contradictory opinions about these variations are so evenly divided that at the same instant buyers expect a rise and sellers a fall." Therefore, not only is the market subject to random external effects, but its own reaction to that news will also to some degree be random.

Furthermore, Bachelier pointed out that the exchange was involved in a kind of narcissistic dance with itself. Because every financial transaction involves a prediction of the future, this means that a speculator on the Bourse cares less about a sober appraisal of an asset's worth than he does about the opinions of his colleagues. His aim is to evaluate how much the market is willing to pay at some time in the future. He doesn't mind overpaying, if he thinks that a greater fool will over-overpay the next day.

Bachelier therefore concluded that movements in the exchange were essentially random. Any connection between causes and effects was too obscure for a human being to comprehend. As he wrote at the beginning of his thesis, "The factors that determine activity on the Exchange are innumerable, with events, current or expected, often bearing no relation to price variation." Mathematical forecasting was therefore impossible. However, he then made a point that underpins much of modern economic theory, which is that one could "establish the laws of probability for price variation that the market at that instant dictates." To accomplish this, he assumed in his calculations that market prices followed the normal distribution, which seemed reasonable given its popularity in the physical sciences.

The theory implied that there could be no GCM, no grand model of the economy that could predict future security prices. The current prices represented a balance between buyers and sellers, and that balance would not shift without some external cause. All changes are therefore due to random external effects—complications—which by definition cannot be predicted. Any foreseeable future event, such as the impact of the seasons on agricultural produce, would be factored into the price. The net expectation for profit of an investor would at any time be zero, because the price of a security was always in balance with its true value, right on the money. The market was a larger version of a Monte Carlo casino. But like a gambler at the casino, an investor could make intelligent bets by figuring the odds and controlling his risk.

All of this went down like a stock-market crash with Poincaré and Bachelier's other supervisors. Poincaré might have discovered chaos, but this did not weaken his faith in the scientist's ability to discern cause and effect. He believed that "what is chance for the ignorant is not chance for the scientists. Chance is only the measure of our ignorance."[15] Bachelier's thesis was awarded an undistinguished grade, which meant that he couldn't find a permanent position for twenty-seven years. And his theory remained out of sight until half a century later, when it stumbled back into town as the random walk theory.

RANDOM, BUT EFFICIENT

Interest in Bachelier's theory revived after a number of studies showed that asset prices did move in an apparently random and unpredictable way—just as he had predicted. In 1953, the statistician Maurice Kendall analyzed movements in stock prices over short time periods and found that the random changes were more significant than any systematic effect, so the data behaved like a "wandering series." In a 1958 paper, the physicist M. F. M. Osborne showed that the proportional changes in a stock's price could be simulated quite well by a random walk—like the drunk searching for his car keys.[16]

This seemed to explain why investors had such difficulty predicting stock movements. In 1933, a wealthy investor called Alfred Cowles III had published a paper showing that the top twenty insurance companies in the United States had demonstrated "no evidence of skill" at picking their investments.[17] If market movements were essentially random, it would be impossible to guess where they were headed.

Bachelier's idea was made manifest in the 1960s by economist Eugene Fama of the University of Chicago in the efficient market hypothesis, or EMH. This proposed that the market consists of "large numbers of rational, profit-maximizers actively competing, with each trying to predict future market values of individual securities."[18] Because any randomness in the market was the result of external events, rather than the activity of investors, the value of a security was always reflected in its current price. There could be no inefficiencies or price anomalies, since these would immediately be detected by investors.

Of course, no market could be totally efficient—especially smaller and less fluid markets such as real estate, where there may be only a small number of buyers interested in a particular property. But to a good approximation, the large bond, stock, and currency markets could be considered efficient. These involve tens of millions of well-informed investors, and they operate relatively free of regulation or restriction. Different versions of the EMH assume varying amounts of efficiency and take into account factors such as insider trading (where traders profit from information that is not widely available). While the EMH is increasingly being debated, as seen below, it still forms the main plank of orthodox economic theory.[19]

Where the EMH view of the market differed from Bachelier's was in the assumption that it was made up of "rational" profit-maximizers. Bachelier had concluded that "events, current or expected, often [bear] no apparent relation to price variation." According to the EMH, however, an efficient market always reacts in the appropriate way to external shocks. If this wasn't the case, then a rational investor would be able to see that the market was over- or under-reacting and profit from the situation. The fact that investors could not reliably predict the market seemed to imply that it behaved like a kind of super-rational being, its collective wisdom emerging automatically from the actions of rational investors.

The EMH naturally posed something of a challenge to economic forecasters: not only was there no way to predict the flow of money using fundamental economic principles, but even the weaker notion of forecasting the movements of individual stocks or bonds seemed out of reach. Nonetheless, individuals, banks, insurers like Prudential, investment firms like Merrill Lynch, large companies, governments, giant financial institutions such as the World Bank and the International Monetary Fund, and perhaps the greatest economic oracle of them all, the U.S. Federal Reserve, which twice a year presents an economic forecast to Congress—all collectively employ thousands of economists who claim to be able to foresee market movements. So what is going on?

MAKING A PROPHET

While mathematical models of physical systems are usually based on the same general principles, economic models vary greatly in both aims and approach. Nowhere is this more true than in academia. The business cycle of expansions and recessions has been modelled based on sunspots, from a Marxist viewpoint, from a Keynesian perspective, as a predator-prey relationship between capitalists and labour, using "real business cycle" theory (which simulates the workforce's reaction to external shocks), and so on. Part of the problem is that unlike colonies of yeast, colonies of humans do not sit still for controlled scientific experiments, so it is hard to prove that any theory is definitely false.

Most economic forecasters fall into one of two camps: the data-driven chartists or the model-driven analysts. Chartists, or technicians, are people who look for recurring patterns in financial records. Perhaps the simplest predictive method is to assume, as the Greeks did, that everything moves in circles, that there is nothing new under the sun. To forecast the future, it suffices to search past records for a time when conditions were similar to today's. Such forecasts often appear in the financial sections of newspapers— when, for example, plots are produced to show that recent stock prices mirror those that preceded a historic crash or fit some pattern that signals the start of a bull or bear market.[20] More sophisticated versions, discussed below, use advanced techniques to detect signals in multiple streams of financial data.

The only problem with chart-following is that, statistically speaking, it doesn't seem to work, at least not for most people. Analysis has shown that, after accounting for the expense of constantly buying and selling securities, chartists on average earn no more money for their clients over the long-term than investors earn for themselves with a naïve buy-and-hold strategy.[21] There are at least two reasons for this. The first is that for the present to perfectly resemble the past, the inflationary environment would have to be the same, interest rates would have to match—in principle, everyone would have to be doing the same thing as before. Which of course doesn't happen. Atmospheric conditions never reproduce themselves exactly, and Frank Knight made a similar statement about the economy in his 1921 work, *Risk, Uncertainty, and Profit*. The economy is not, therefore, constrained to follow past behaviour.

The second reason is that if a genuine pattern emerges, it is only a matter of time before investors notice it, at which point it tends to disappear. Suppose, for example, that investors have a habit of selling stocks at year end in order to write off the losses on their taxes. The price of stocks should then dip, rebounding in January. While the January effect, as it became known, may once have been real, if rather subtle, it became much harder to detect after the publication of a book called *The Incredible January Effect.*[22] Everyone started buying in January to take advantage of it, so naturally prices went up and any anomaly disappeared. Like a biological organism, the economy evolves in such a way that it becomes less predictable.

The fascination with financial charts often says more about the human desire for order than it does about the markets themselves. Any series of numbers, even a random one, will begin to reveal patterns if you look at it long enough. Indeed, in an efficient market, price movements would be random and any pattern no more than an illusion. As Eugene Fama put it, "If the random walk model is a valid description of reality, the work of the chartist, like that of the astrologer, is of no real value in stock market analysis."[23]

Unlike chartists, fundamental analysts base their stock choices on their estimate of a stock's "intrinsic value."[24] This relies on a prediction of the company's future prospects and dividends, as well as the effects of volatility, inflation, and interest rates. If the forecaster's estimate is higher than the market price, then he predicts the stock will rise; if his estimate is lower, it should fall. The best-known proponent of this approach is Warren Buffet, known as the "Oracle of Omaha" for his canny stock picks.

The challenge of this approach is that it requires predictions of the future that are better than those the market is making. This in turn often involves some form of chart reading. Suppose a company is set up to market a new product. New companies, products, or innovations that have gone on to be successful often follow an S-shaped curve like that shown in figure 6.1.[25] Starting from a low level, sales grow exponentially as word of mouth spreads and the idea catches on. Success begets success in a positive feedback loop. The company or product then enters a period of steady and sustainable growth. But nothing can grow forever, and eventually the growth will saturate.

In principle, you could predict a company's future prospects if you knew where it was on this curve. Unfortunately, this is impossible: the company could be snuffed out at day one, or it could turn into the next Microsoft. Also, the S-shaped curve is not the only possibility. The company might shoot up, then shoot back down when its product goes out of fashion, then make a startling comeback selling something else (Apple). The market itself might change or even collapse. Most companies that were huge a hundred years ago no longer exist, because the demand for typewriters and horse carriages is not what it was. Estimates of future growth are therefore highly uncertain, especially over the long term. They are really a judgment on how compelling a particular investment story is.

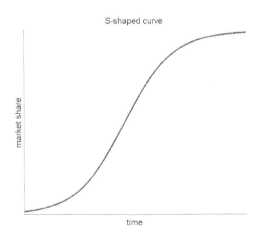

FIGURE 6.1. The S-shaped development curve goes through three stages: an initial stage of exponential growth, a period of steady growth, and finally a period of saturation. This plot shows market share of a hypothetical company as a function of time.

Once again, the future need not resemble the past. In fact, the record of most analysts is not much better than that of chartists. Their predictions routinely fail to beat naïve forecasts, and funds that hold all the stocks in a particular index routinely outperform managed funds, at least after expenses. Some do much better, but the occasional success story may be simple luck: every investment strategy has to win some of the time. As the economist Burton G. Malkiel wrote, "Financial forecasting appears to be a science that makes astrology look respectable."[26]

The mediocre performance of most stock pickers seems to confirm the EMH: in an efficient market, stock selection should be a waste of time because the true value of any asset is always reflected in the price. It is impossible to make better forecasts than the market itself, at least on a consistent basis. Price fluctuations are a random walk, so they are inherently unpredictable. According to Fama, "If the analyst has neither better insights nor new information, he may as well forget about fundamental analysis and choose securities by some random selection procedure."[27] Throwing darts at the financial section of the *Wall Street Journal* is said to work quite well as a stock-picking technique.

Prices of basic commodities such as oil are especially difficult to forecast, because both supply and demand are subject to shifting political, economic, and geographic factors.[28] The export quotas of OPEC countries are determined by their oil reserves, so there is an incentive to inflate the estimates. Many of the OPEC nations involved are notoriously unstable (the Middle East), unpredictable (Venezuela), or at risk of terrorism. As much as a fifth of the 2006 oil price is made up of the so-called political-risk premium, which spikes every time there is a perceived threat against an oil supplier or the transport network. Hurricanes are also a factor, as seen by the fluctuations in oil price as the storms duck and weave their way towards Gulf coast refineries.

Even markets that are not that efficient are hard to predict—like real estate, where there are few buyers per property, and the process of buying or selling is relatively slow and expensive. Some housing markets are believed to follow cyclical trends.[29] When house prices are at a low level relative to the cost of renting, new buyers enter the market because it is affordable. As prices begin to rise, speculators join the party, driving the price up with positive feedback. When house prices grow too high relative to rents, the supply of first-time buyers is cut off and the market reaches a plateau—negative feedback. If prices begin to fall, more sellers will try to off-load their properties, driving prices down: positive feedback in the other direction. The cycle therefore continues. However, there are many other factors to take into account, and even if such a cycle does exist, the pattern constantly varies. It is hard to know when turning points will occur, and even harder to beat a buy-and-hold strategy once transaction fees are taken into account.

If no individual person can predict the future, perhaps several people can. In 1948, the RAND Corporation came up with a method of making decisions based on an ensemble approach. A number of experts were polled on a series of questions. The questions were refined based on their input, and the process repeated until a group consensus was obtained. The method, known as Delphi after the Greek oracle, was first used by the United States Defense Department to investigate what would happen in a nuclear war, but it was soon adopted by businesses for making financial decisions.

Imagine that you are a *theoprope* who has time-travelled from ancient Greece. You have heard of this place called Delphi. You show up at an office in a high-rise in downtown New York. You hand the bemused-looking secretary a goat. She shows you to a room. Inside are . . . a group of management consultants. "Hello, Mr.

Theopropous," says one. "We've come to a consensus. We will have the Mediterranean lunch special, all round." You run out screaming.

Not only did the new-version Delphi lack a certain mystique, but it wasn't very good at predicting the future. A 1991 study by the experimental psychologist Fred Woudenberg showed that the Delphi was no more accurate than other decision-making methods, because "consensus is achieved mainly by group pressure to conformity."[30]

So if financial analysts of all stripes cannot predict the future, why are there so many of them, and why are they so well reimbursed? Would the market work just as well without them? The fact is that the market augurs exist because they do make a lot of money—for themselves and their employers. Buying and holding may be good for the client, but the fastest way to generate commissions is by buying, selling, buying again, and so on. Also, there is a strong market for predictions, and accuracy is of secondary importance. It is always easy to generate data that makes it look as if a method has been highly reliable in the past. Finally, as discussed below, some stock pickers really do manage to beat the market, at least for a while.

Of course, if all the chartists, analysts, and consultants were replaced with chimps armed with darts, the market would disintegrate pretty rapidly. If the market has any semblance of efficiency, it is because of the combined efforts of predictors who make up the investor ecosystem. You could argue that if the system has parasites, it is those who invest in index funds. These funds tend to purchase the stocks that have been picked by active managers, so investors benefit passively from their decisions.

THE GLOBAL CAPITAL MODEL

Chartists and analysts both focus on particular assets or asset classes. On a grander scale, major private and governmental banks, some economic-forecasting firms, and institutions such as the Organization for Economic Co-operation and Development (OECD) have developed large econometric models that attempt, in the style of Jevons, to simulate the entire economy by aggregating over individuals. Their aim is to make macro-economic forecasts of quantities such as gross domestic product (GDP), which is a measure of total economic output, and to predict recessions and other turning points in the economy, which are of vital interest to companies or governments. The models are similar in principle to those used in weather forecasting or biology, but they involve hundreds or sometimes thousands of economic variables, including tax rates, employment, spending, measures of consumer confidence, and so on.

In these models, like the others, the variables interact in complex ways with multiple feedback loops. An increase in immigration may cause a temporary rise in unemployment, but over time, immigration will grow the economy and create new jobs, so unemployment actually falls. This may attract more immigrants, in a positive feedback loop, or heighten social resistance to immigration from those already there—negative feedback. The net effect depends on a myriad of local details, such as what each immigrant actually does when he arrives. The model equations therefore represent parameterizations of the underlying complex processes, and they attempt to capture correlations between variables, either measured or inferred from theory. Again, the combination of positive and negative feedback loops tends to make the equations sensitive to changes in parameterization. The model is checked by running it against historical data from the past couple of decades and comparing its predictions with actual results. The parameters are then adjusted to improve the performance, and the process is repeated until the model is reasonably consistent with the historical data.

While the model can be adjusted to predict the past quite well—there is no shortage of knobs to adjust—this doesn't mean that it can predict the future. As an example, the black circles in figure 6.2 show annual growth in GDP for the G7 countries (the United States, Japan, Germany, France, Italy, the United Kingdom, and Canada). The white circles are the OECD forecasts, made a year in advance, and represent a combination of model output and the subjective judgment of the OECD secretariat. The forecast errors, which have standard deviation 0.95, are comparable in magnitude to the fluctuations in what is being forecast, with standard deviation 1.0. The situation is analogous to the naïve "climatology" forecast in weather prediction, where the forecast error is exactly equal to the natural fluctuations of the weather.

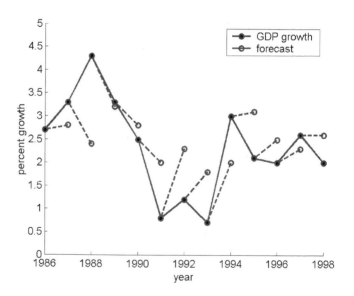

FIGURE 6.2. GDP growth for the G7 countries, plotted against the OECD one-year predictions, for the period 1986 to 1998. Standard deviation of errors is 0.95, that of the GDP growth is 1.0.[31]

These results are not unique to the OECD, but are typical of the performance of such forecasts, which routinely fail to anticipate turning points in the economy.[32] As *The Economist* noted in 1991, "The failure of virtually every forecaster to predict the recent recessions in America has generated yet more skepticism about the value of economic forecasts."[33] And again in 2005, "Despite containing hundreds of equations, models are notoriously bad at predicting recessions."[34] In fact, if investors used econometric models to predict the prices of assets and gamble on the stock or currency markets, they would actually lose money.[35] Consensus between an ensemble of different models is no guarantor of accuracy: economic models agree with one another far more often than they do with the real economy.[36] Nor does increasing the size and complexity of the model make results any better: large models do no better than small ones.[37] The reason is that the more parameters used, the harder it is to find their right value. As the physicist Joe McCauley put it, "The models are too complicated and based on too few good ideas and too many unknown parameters to be very useful."[38] Of course, the parameters can be adjusted and epicycles added (just as the ancients did with the Greek Circle Model) until the model agrees with historical data. But the economy's past is no guide to its future.

Similar models are used to estimate the impact of policy changes such as interest hikes or tax changes—but again, the results are sensitive to the choices of the modeller and are prone to error. In one 1998 study, economist Ross McKitrick ran two simulations of how Canada's economy would respond to an average tax cut of 2 percent, with subtle differences in parameterization. One implied that the government would have to cut spending by 27.7 percent, while the other implied a cut of only 5.6 percent—a difference of almost a factor five. In the 1990s, models were widely used to assess the economic effects of the

North American Free Trade Agreement (NAFTA). A 2005 study by Timothy Kehoe, however, showed that "the models drastically underestimated the impact of NAFTA on North American trade, which has exploded over the past decade."

In a way, the poor success rate of economic forecasting again seems to confirm the hypothesis that markets are efficient. As Burton Malkiel argued, "The fact that no one, or no technique, can consistently predict the future represents . . . a resounding confirmation of the random-walk approach."[39] It also raises the question why legions of highly paid professionals—including a large proportion of mathematics graduates—are employed to chart the future course of the economy. And why governments and businesses would follow their advice.

RATIONAL ECONOMISTS

Now, a non-economist might read the above and ask, in an objective, rational way, "Can modern economic theory be based on the idea that I, and the people in my immediate family, are rational investors? Ha! What about that piece of land I inherited in a Florida swamp?" Indeed, the EMH sounds like a theory concocted by extremely sober economists whose idea of "irrational behaviour" would be to order an extra scoop of ice cream on their pie at the MIT cafeteria. Much of its appeal, however, lay in the fact that it provided useful tools to assess risk. As Bachelier pointed out, the fact that the markets are unpredictable does not mean that we cannot calculate risks or make wise investments. A roll of the dice is random, but a good gambler can still know the odds. The EMH made possible a whole range of sophisticated probabilistic financial techniques that are still taught and used today. These include the capital asset pricing model, modern portfolio theory, and the Black-Scholes formula for pricing options.

The capital asset pricing model was introduced by the American economist William F. Sharpe in the 1960s as a way to value a financial asset by taking into account factors such as the asset's risk, as measured by the standard deviation of past price fluctuations. It provides a kind of gold standard for value investors. The aim of modern portfolio theory, developed by Harry Markowitz, was to engineer a portfolio that would control the total amount of risk. It showed that portfolio volatility can be reduced by diversifying into holdings that have little correlation. (In other words, don't put all your eggs in the same basket, or even similar baskets.) Each security is assigned a number β, which describes its correlation with the market as a whole. A β of 1 implies that the asset fluctuates with the rest of the market, but a β of 2 means its swings tend to be twice as large and a β of 0.5 means it is half as volatile. A portfolio with securities that tend to react in different ways to a given event will result in less overall volatility.

The Black-Scholes method is a clever technique for pricing options (which are financial instruments that allow investors to buy or sell a security for a fixed price at some time in the future). Aristotle's *Politics* describes how the philosopher Thales predicted, on the basis of astrology, that the coming harvest would produce a bumper olive crop. He took out an option with the local olive pressers to guarantee the use of their presses at the usual rate. "Then the time of the olive-harvest came, and as there was a sudden and simultaneous demand for oil-presses he hired them out at any price he liked to ask. He made a lot of money, and so demonstrated that it is easy for philosophers to become rich, if they want to; but that is not their object in life."[40] Today, there are a wide variety of financial derivatives that businesses and investors use to reduce risk or make a profit. Despite the fact that options have been around a long time, it seems that no one, even philosophers, really knew how to price them until Black-Scholes. So that was a good thing.

All of these methods were built on the foundations of the EMH, so they treated investors as inert and rational "profit-maximizers," modelled price fluctuations with the bell curve, and reduced the measurement of risk to simple parameters like volatility. There were some objections to this rather sterile vision of the economy. Volatility of assets seemed to be larger than expected from the EMH.[41] Some psychologists even made the point that not all investors are rational, and they are often influenced by what other investors are doing. As Keynes had argued in the 1930s, events such as the Great Depression or the South Sea Bubble could be attributed to alternating waves of elation or depression on the part of investors. The *homme moyen,* it was rumoured, was subject to wild mood swings.

On the whole, though, the EMH seemed to put economic theory on some kind of logical footing, and it enabled economists to price options and quantify risk in a way that previously hadn't been possible. The "model" for predicting an asset's correct value was just the price as set by the market, and it was always perfect. Even if all investors were not 100 percent rational, the new computer systems that had been set up to manage large portfolios had none of their psychological issues. Perhaps for the first time, market movements could be understood and risk contained. To many economists, the assumptions behind the orthodox theory seemed reasonable, at least until October 19, 1987.

COMPLICATIONS

According to random walk theory, market fluctuations are like a toss of the die in a casino. On Black Monday, the *homme moyen,* Mr. Average, sat down at a craps table. The only shooter, he tossed two sixes, a loss. Then two more, and two more. Beginning to enjoy himself in a perverse kind of way—nothing so out of the normal had happened to him in his life—he tried several more rolls, each one a pair of sixes. People started to gather around and bet that his shooting streak would not continue. Surveillance cameras in the casino swivelled around to monitor the table. The sixes kept coming. Soon, the average man was a star, a shooting star flaming out in a steady stream of sixes and taking everyone with him. At the end of the evening, when security guards pried the die out of his fingers, he had rolled thirty twelves in a row and was ready for more. The net worth of everyone in the room who bet against his streak had decreased, on average, by 29.2 percent. Someone in the house did the math and figured out the odds of that happening were one in about ten followed by forty-five zeros—math-speak for impossible.

Black Monday, when the Dow Jones index fell by just that amount, was an equally unlikely event—and a huge wakeup call to the economics establishment. According to the EMH, which assumes that market events follow a normal distribution, it simply shouldn't have happened. Some have theorized that it was triggered by automatic computer orders, which created a cascade of selling. Yet this didn't explain why world markets that did not have automatic sell orders also fell sharply. Unlike the South Sea Bubble, which was at least partly rooted in fraud, Black Monday came out of nowhere and spread around the world like a contagious disease. It was as if the stock market suddenly just broke. But it has been followed by a string of similar crashes, including one in 1998 that reduced the value of East Asian stock markets by $2 trillion, and the collapse of the Internet bubble. Perhaps markets aren't so efficient or rational after all.

A strong critic of efficient-market theory has been Warren Buffet, who in 1988 observed that despite events such as Black Monday, most economists seemed set on defending the EMH at all costs: "Apparently, a reluctance to recant, and thereby to demystify the priesthood, is not limited to theologians." Another critic was an early supporter (and Eugene Fama's ex-supervisor), the mathematician Benoit Mandelbrot. He is best known for his work in fractals (a name he derived from the Latin *fractus,* for "broken"). Fractal geometry is a geometry of crooked lines that twist and weave in unpredictable ways. Mandelbrot turned the tools of fractal analysis to economic time series. Rather than being a random walk, with each change following a normal distribution, they turned out to have some intriguing features. They had the property, common to fractal systems, of being self-similar over different scales: a plot of the market movements had a similar appearance whether viewed over time periods of days or years (see boxed text below). They also had a kind of memory. A large change one day increased the chance of a large change the next, and long stretches where little happened would be followed by bursts of intense volatility. The markets were not like a calm sea, with a constant succession of "normal" waves, but like an unpredictable ocean with many violent storms lurking over the horizon.

BORDERLINE NORMAL

Lewis Fry Richardson, the inventor of numerical weather forecasting, once did an experiment in which he compared the lengths of borders between countries, as measured by each country. For example, Portugal believed its border with Spain was 1,214 kilometres long, but Spain thought it was only 987 kilometres. The problem was that the border was not a straight line, so the length would depend on the scale of the map used to measure it. A large scale includes all the zigs and zags, while a small scale misses these and give a shorter result. If the measured length is plotted as a function of the scale, it turns out to follow a simple pattern known as a power law. In a 1967 paper, Benoit Mandelbrot showed how this could be used to define the border's fractal dimension, which was a measure of its roughness.

If the border was a one-dimensional straight line, its length on the map would vary linearly with the scale—a map with twice the scale would show everything twice as long, including the border. The area of a two-dimensional object such as a circle would vary with the scale to the power of two—double the scale and the area increases by a factor of four. If the length of a border increases with scale to the power D, then D plays the role of dimension. Mandelbrot showed that the British coastline has a fractal dimension D of about 1.25.

Like clouds, fractal systems reveal a similar amount of detail over a large range of scales. There is no unique "normal," or correct, scale by which to measure them. Similarly, the fluctuations of an asset or market show fractal-like structure over different time scales, which makes analysis using orthodox techniques difficult.

In the 1990s, researchers tested the orthodox theory by poring over scads of financial data from around the world. The theory assumes, for example, that a security has a certain volatility, and that it varies in a fixed way with other assets and the rest of the market. The volatility can in principle be found by plotting the asset's price changes and calculating the standard deviation. In reality, though, the actual distributions have so-called fat tails, which means that extreme events—those in the tails of the distribution—occur much more frequently than they should. One consequence is that the volatility changes with time.[42] An asset's correlation with the rest of the market is also not well defined. It is always possible to plot two data sets against each other and draw a straight line through the resulting cloud of points, as Galton did for his height measurements. But economic data is often so noisy that the slope of the line says little about any underlying connection between sets.[43]

Perhaps the biggest problem with the orthodox economic theory, though, is its use of the bell curve to describe variation in financial quantities. When scientists try to model a complex system, they begin by looking for symmetric, invariant principles: the circles and squares of classical geometry; Newton's law of gravity. Einstein developed his theory of relativity by arguing that the laws of physics should remain invariant under a change of reference frame. The normal distribution has the same kind of properties. It is symmetric around the mean, and is invariant both to basic mathematical operations and to small changes to the sample. If the heights of men and women each follow a bell curve, then the mid-heights of couples will be another bell curve, as in figure 5.1 (see page 178). Adding a few more couples to the sample should not drastically change the average or standard deviation.

The normal distribution means that volatility, risk, or variation of any kind can be expressed as a single number (the standard deviation), just as a circle can be described by its radius or a square by the length of one side. However, the empirical fact that asset volatility changes with time indicates that this is an oversimplification. The bell curve can be mathematically justified only if each event that contributes to fluctuation is independent and identically distributed. But fluctuations in the marketplace are caused by the decisions of individual investors, who are part of a social network. Investors are not independent or identical, and therefore they cannot be assumed to be "normal." And as a result, neither can the market. Indeed, it turns out that much of the data of interest is better represented by a rather different distribution, known as a power-law distribution.

POWER TO THE PEOPLE

Suppose there existed a country in which the size of its cities was normally distributed, with an average size of half a million. Most people would live in a city that was close to the average. The chances of any city being either smaller or larger than average would be roughly the same, and none would be extremely large or extremely small. The expressions "small town" and "big city" would refer only to subtle variations. The pattern in real countries is quite different. One 1997 study tabulated the sizes of the 2,400 largest cities in the United States. The study's authors found that the number of cities of a particular size varies inversely with the size squared (to the power of two). This so-called power-law pattern continues from the largest city, New York, right down to towns of only 10,000 residents.[44] This distribution is highly asymmetrical. For each city with a certain population, there are, on average, four cities of half the size, but only a quarter the number of cities twice the size—so there are many more small towns than would be predicted from a normal distribution. The distribution is fat-tailed—the largest metropolis, New York, is far bigger than the mean. Even the home of Wall Street, it seems, is not normal. The same pattern was found for the 2,700 largest cities in the world.

A similar asymmetrical distribution—sometimes known as Pareto's law, after the nineteenth-century Italian who first discovered it—holds for personal wealth. Perhaps Warren Buffet so distrusted the EMH because if the normal distribution were correct, he shouldn't exist. Most people would enjoy a uniform degree of financial success; no one would be extremely poor or exceptionally rich. In reality, however, the wealthy are arranged like cities. For every millionaire in the United States, there are about four people with half a million, sixteen with a quarter million, and so on. If heights were arranged in a similar way, then most people would be midgets, but a small number would be giants.

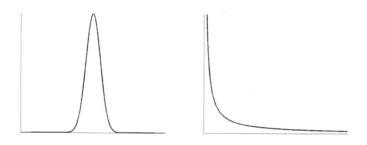

FIGURE 6.3. The left panel shows a normal distribution, the right a power-law distribution. The horizontal axis could, for example, represent wealth or city size. In reality, the power-law distribution holds only over a certain range, so there are cutoffs in both the vertical and the horizontal axes. In the normal distribution, it is possible to assume that most samples are close to the mean. In a power-law distribution, we cannot ignore the samples in the extreme right-hand tail (because they are the most important in terms of the quantity measured) or those in the left (because there are so many of them).

The growth of an investor's assets, or a city's population, is not a completely random event, but depends on its position in a connected network.[45] A person thinking to move is more likely to know people or employers in a large town than in a small town. Big cities draw people to them like moths. Money attracts more money; the rich get richer; strong countries attract investment, while those on the periphery remain vulnerable. The economy consists of both negative feedback loops, like Adam Smith's invisible hand, and positive feedback loops, which amplify differences. In such situations, Galton's principle of "regression to the mean" is highly misleading. What goes up needn't come down.

Imagine that several different restaurants of similar quality start business on the same day, and that, for some random reason, one restaurant initially attracts a few more customers than the others. (Perhaps its location is slightly better, or the proprietor is well known and popular.) A visitor who lacks detailed information will likely choose the restaurant that holds a few more people, just because that implies that it must be reasonably good. Therefore, the number of customers is amplified by positive feedback: the more customers there are, the more new ones will be generated. As business improves, the proprietor may invest the proceeds in better ingredients (product quality), in advertising (brand equity), or in expanding the restaurant (economies of scale). The restaurant pulls further ahead of its competition. Growth may eventually saturate—when, for example, some of the customers grow bored with the menu. The company therefore follows the S-shaped curve of figure 6.1 (see page 237). Meanwhile, the restaurant down the road, which was just as good to start off with, quietly goes bust within a year or two—as most new businesses do. Positive feedback amplifies changes, be they good or bad, so it can lead to collapse as well as growth.

The power-law distribution holds, in an approximate fashion, over many different phenomena, from the size of earthquakes to the number of interactions in a biological network (like that in figure

5.4 on page 203).[46] Unfortunately, this does not help much with prediction of particular events, because it implies that there is no typical representative. It is no longer possible to predict that an individual sample will be close to the mean, give or take a standard deviation or two. Even the calculation of the mean depends critically on the sample. It has been estimated that if Bill Gates attends a baseball game at Safeco Field in Seattle, the average net worth of those also in attendance increases by a factor of four.[47] Leave him out of the calculation, and you miss the most important person. There is no easy mathematical shortcut. Details matter, and nothing is "normal."

Like customers choosing a restaurant, investors must be viewed as individual but interdependent agents operating in a highly connected network. The decisions they make can ripple through the entire system. Any accurate model of the economy would have to take into account these intricate social dynamics, which are explicitly ignored by the EMH. Despite these drawbacks, no serious alternative is close to dislodging the orthodox theory from the throne of predictive economics. So why is this the case? And is there anything better available?

STICKY THEORY

The orthodox economic theory has proved so resilient because it is both hard to beat—in the sense that other equations do little better at predicting the future or estimating risk—and highly adaptable. More elaborate versions of the basic theory attempt to correct the flaws while retaining the same structure. The field of behavioural economics, for example, addresses psychological effects such as loss aversion. Owners of financial assets hate to sell at a lower price than they paid, so they resist selling after a downturn. Therefore, prices tend to be "sticky" on the way down. Models can be tweaked and adjusted to accommodate these behavioural effects, by incorporating parameterizations of investor psychology, which still treat each investor as having fixed tastes and preferences. These adjustments relax the condition of investor rationality but still assume that investors can be modelled rationally. They therefore do not address the underlying problem, which is that the economy is a social process that cannot be reduced to law.[48]

Another approach is to stretch the theory so it better allows for extreme events. Consider, for example, the Mexican peso crisis of 1994–95. The standard deviation of the peso-dollar exchange rate from November 19, 1993, to December 16, 1994, was 0.47 percent. According to random walk theory, someone buying pesos on December 19, 1994, should have expected to lose no more than 3.5 percent of his dollar investment over a two-week period, ninety-nine times out of a hundred. As it turns out, losses were 65 percent, which in principle should not have happened in a billion years.[49] The reasons behind the high losses—a peasant uprising in Chiapas and the death of a presidential candidate—were local to Mexico. But the crisis was soon followed by sister crises around the world: the collapse of Asian currencies (1997–98) and Russian bonds (1998).

To account for such sudden changes in volatility, more sophisticated risk-assessment techniques have been developed, such as the generalized autoregressive conditional heteroskedasticity (GARCH).[50] The last word, to you and me, means changing variability; autoregressive means that the changes depend on past behaviour. In its most basic form, GARCH produces a distribution of price changes that better accounts for extreme events, but at the expense of two new parameters. These add to the number of unknowns, as well as the number of Greek letters in the equations, and still cannot begin to account for the inherently social and political nature of the market (there is no parameter for peasant uprising). As Mandelbrot put it, "The high priests of modern financial theory keep moving the target. As each anomaly is reported, a 'fix' is made to accommodate it. . . . But such ad hoc fixes are medieval. They are akin to the countless adjustments that defenders of the old Ptolemaic cosmology made to accommodate pesky new astronomical observations."[51] In practice, it seems hard to find numerical techniques that improve greatly on the orthodox methods, which is why they are still orthodox. Forecasters can always try to get around shortcomings in the model output by applying their subjective judgment. One currency trader said that pricing techniques such as Black-Scholes only represent the numerical approach to the problem: "Although used as a first cut, the actual price quoted takes non-quantitative factors into account."[52]

Another reason for the theory's endurance is the lock-in effect: it has become entrenched in academic circles and elsewhere. Just as the perfect model hypothesis in weather forecasting enables researchers to do complicated, pseudo-probabilistic analyses of the atmosphere, so orthodox theory allows the development of intimidating and authoritative financial techniques. As the economist Paul Ormerod wrote, "Maximizing behaviour, for all its faults, is a valuable security blanket for many economists. It enables the mathematics of differential calculus to be applied to their theories, and for some intellectually satisfying results to be obtained. It has the side-benefit, too, of inducing mortal terror in many scholars from other disciplines in the social sciences who lack the required amount of mathematical training."[53] This is reminiscent of Iamblichus's statement about the Pythagoreans: "Their writings and all the books which they published were not composed in a popular and vulgar diction, so as to be immediately understood, but in such a way as to conceal, after an arcane mode, divine mysteries from the uninitiated."[54]

One defence of the orthodox theory is that it correctly predicts that the market is unpredictable. If, it is argued, the market is irrational, then a rational investor would be able to consistently outsmart it.[55] But while an efficient market implies unpredictability, the opposite is not true. It is actually rather strange that unpredictability is cited as evidence of logical calm. A drunken man's stumbling (the original inspiration for the random walk) might be unpredictable, but we wouldn't call his behaviour hyper-rational.

Perhaps the greatest attraction of orthodox theory lies in its subtext. It allows economists to maintain the illusion that the economy is fundamentally rational, and that their models are correct. The causes of error are externalized to random external shocks. The tenets of the theory—that investors are independent and fixed in their tastes, and can look into the future to calculate value—reflect the Pythagorean ideals of an ordered universe. They are also exactly the assumptions that need to be made if economic models are to be considered accurate and economists are to maintain any oracular authority. As soon as we admit that the economy is a complex set of interactions in a huge connected network, and that it involves individuals whose "beliefs, interpretations and justifications evolve and transform themselves continuously," then the idea of accurate mathematical models begins to seem, to paraphrase Immanuel Kant, a little absurd. The system is uncomputable, and there is no Apollo's arrow to fly into its future. It is interesting that Bachelier, whose work was at first spurned, was not allowed back in the fold until his random walk hypothesis was reconciled with this image of a rational marketplace.

Indeed, a huge amount of effort has gone into explaining rationally why the market is rational. This leads to the kind of double-think that allowed astronomers to retain the notion of perfectly circular motion, despite much evidence to the contrary, for about 2,000 years.[56] Another example of double-think occurred during the summer of 1988, when a severe drought in the United States

Midwest affected the supply of corn and soybeans. As time went on and the drought continued, prices of these commodities rose to extreme heights. Then one day, Chicago experienced a tiny, insignificant amount of rain—one imagines a few drops sprinkled on the balding pate of a broker as he headed out to work in the morning—and prices collapsed. Some saw this as a sign that markets might be a touch on the manic, oversensitive side, but efficient-market enthusiasts disagreed. One top economist insisted that the response in the market was quite rational, because investors know that weather tends to be persistent, so they all updated their forecasts with this new information and concluded that the drought would not continue.[57] Given the uncertainty in weather forecasts of any type, it seems to be stretching the meaning of the word "rational" to allow a few rain clouds to so drastically change expectations for the future.

THE PSYCHOLOGY OF ECONOMICS

Here are some factors that behavioural psychologists believe affect investors:

COMPARTMENTALIZING. Investors divide problems into parts and treat each separately, instead of looking at the big picture. Losing ten dollars on the street feels worse than losing the same amount in a stock portfolio because each event is handled in a different mental compartment.

TREND FOLLOWING. Extrapolating from current conditions tends to fuel bubbles because when the market is going down, people expect it to stay down, and when it is going up, the sky is the limit.

LOSS AVERSION. People take less pleasure in winning ten dollars than they do pain in losing the same amount. A consequence is that investors will avoid selling assets at a loss.

DENIAL. Investors maintain beliefs even if they are at odds with the evidence. This results in cognitive dissonance.

SUGGESTION. They are overly influenced by the opinions of others.

STATUS QUO BIAS. They tend to avoid change. It means that transactions, which always involve some kind of change, have a hidden extra cost.

ILLUSORY CORRELATIONS. Investors look for patterns where they don't exist. This is also called superstition.

Interestingly, such psychological effects—denial, the power of suggestion, status quo bias, and fear of loss—may also explain why human beings cling to their predictive models even when the results don't agree with reality.

INVESTMENT STORIES

Of course not all economists, and certainly not all traders, agree with the hypothesis that the market is without even pockets of predictability. People such as the former hedge fund manager George Soros do quite well by following their intuitive feel for market sentiment, and betting against trends. It is interesting to compare the EMH vision of rational investors with Soros's notion of "radical fallibility," which contends that "all the constructs of the human mind . . . are deficient in one way or another."[58] (This includes investment ideas or stories—such as South Sea gold, Internet stocks, and indeed abstract economic theories like the EMH—which grow and reinforce themselves in a positive feedback loop as they become established in the investment community.) Profit opportunities arise when you can spot the flaw in the story and bet against it. Because the market is constantly changing and adapting, investment strategies and mental models are themselves nothing but "fertile fallacies" that must be fixed or discarded when they no longer work.

The dangers of excessive confidence in models were amply illustrated by the 1998 collapse of Long-Term Capital Management LP. This hedge fund had a number of economics luminaries on its ticket, including Myron Scholes (of the Black-Scholes formula). It used efficient-market theory to construct complicated and highly leveraged financial bets, which worked well until August 1998, when the Russian government decided to throw efficiency to the winds and default on its bonds. The subsequent market collapse, unanticipated by the equations, meant that to avoid an even greater crisis, the firm had to be rescued in a $3.6-billion bailout.

Attempts are still made to develop investment techniques based on quantitative, predictive models, but they tend to be data-driven rather than model-driven. These can involve classical statistical techniques, which search for correlations in data, or biology-inspired techniques involving neural networks and genetic algorithms. The first biology-inspired approach simulates the way that the brain works by setting up a network of artificial "neurons" that learn to detect patterns in streams of financial data. The latter approach sets alternative algorithms into competition, then chooses the winner in a process akin to natural selection. The aim is not to simulate the underlying economy, but to seek trends in the financial data itself. Such methods may analyze anything at all—say, the dollar-sterling exchange rate and the price of pork bellies—to detect patterns that (for good reason) elude most investors. In the world of finance, even a tiny advantage can lead to substantial profits—at least if you are backed by a large bank, so that positions can be heavily leveraged and transaction costs controlled. The Prediction Company, set up by physicists and funded by the Swiss bank UBS AG, is one firm which uses such techniques.[59] It is hard to judge the firm's success in the somewhat opaque world of financial prediction, but it is still around after more than fifteen years.

Ultimately, though, all such models, like those of the chartist, depend on the future's resembling the past. They seem more alchemy than science. One trader at a major New York investment firm, which employs a large team of about a hundred "quants," or quantitative analysts, told me that it was a "matter of debate" within the firm whether research into predictive mathematical models, beyond simple valuation tools, was worthwhile for them. This is also why the chairman of the U.S. Federal Reserve has yet to be replaced by a machine. The market is a social process that deals in fictions as well as reality, words as well as numbers. Mathematical models certainly have their uses, but they only give part of the story and can be misleading if taken out of context. In the markets, number is not all. Just as computers are not good at spotting social trends or interpreting stories, they do not seem brilliant at making market predictions. For most investors, the teachings of the EMH are quite useful—diversify among different asset classes, avoid expensive management fees, and be wary of prophets who claim to foretell the future. To which could be added: risks may be larger than they appear.

THE LIVING ECONOMY

Even if models are not predictive, they can still be used to capture aspects of the economy's dynamics in a more realistic way. Models have been produced of a hypothetical stock market, for example, in which individual agents are assumed to have their own preferred investment strategy.[60] Some will tend to follow the current psychology and accentuate market swings (positive feedback), while others base their decisions on value analysis and go against trend when they feel the market is too high or too low (negative feedback). The price at any time represents a dynamic balance between these forces: subjective greed or panic versus "objective" calculations of value. The behaviour gets interesting when the market participants are allowed to influence one another: if a majority thinks that the market is going to tank, then this mood eventually deflates the most optimistic chart-follower. Predictions therefore affect the future in a self-reinforcing, positive feedback loop. Tastes and preferences are treated as dynamic rather than fixed. The result is an inherently unstable system that mimics typical market behaviour, including booms and busts. There is no invisible hand to guide it to equilibrium; no such equilibrium even exists.

Agent-based models are an interesting area of research activity, and they can yield insight into the underlying dynamics of markets, as well as expose flaws in the orthodox theory. They can also be used to do "what if" simulations for particular situations—for example, to see how a change in a company's transportation network might affect delivery times. However, it is questionable whether they can help to make more general predictions. It is one thing to produce market-like behaviour in an abstract system, another to forecast what will happen in the real market.[61]

One phenomenon poorly explained by orthodox economic theory is the apparent excess volatility of financial assets, which is inconsistent with the hypothesis that volatility is caused purely by external shocks. Comparisons with biological systems suggest reasons why this is the case. The metabolism of a cell is controlled by many different proteins. The number of molecules of each protein varies in a random way, owing to stochastic effects. The cell also misreads cues from the environment. The resulting variation can be interpreted as a kind of prediction error: the cell produces too many of some proteins and not enough of another, and it also reacts to its own errors. The result is that even in the absence of external shocks, the rate of metabolism will fluctuate randomly. Organisms, like the yeast in Chapter 5, have evolved complex feedback loops that reduce but do not eliminate this natural variation.

Similarly, an economy can be seen as a kind of super-organism and its output a measure of metabolism. The economy takes in resources from people and the environment, transfers them into material wealth, and expels waste products. Value circulates through the economy, changing form from matter to money to labour in a ceaseless flow. The net output is controlled by a large number of individuals and companies, each of which is constantly adjusting output in accord with its predictions about the future. Because the predictions are erroneous, control is incomplete, and the response of each actor is inherently creative, net output will fluctuate randomly of its own accord, adding to volatility.[62] The market is therefore not a dead thing, at equilibrium; as traders like to say, it has a life of its own.

Just as biological systems dislike excess volatility and incorporate complex feedback loops to reduce it, sophisticated economies develop mechanisms to damp out volatile fluctuations. An example is insurance. Let's say a disaster such as a crop-destroying drought has driven individual farmers, whose weather predictions were wrong, into bankruptcy. This has knock-on effects on the rest of the economy, which amplifies the disaster's impact through positive feedback. But with insurance, the loss is dissipated through financial structures that are less vulnerable to collapse than an individual farmer—negative feedback. Similarly, tools such as options or futures contracts allow businesses to smooth out fluctuations in prices and increase their own value. These effects can be simulated (but not precisely predicted) using simple top-down models that capture overall function.

The biological analogy suggests that while the orthodox theory is right to state that the market is practically impossible to predict, it is right for the wrong reasons. The EMH assumes that market fluctuations are the result of random external shocks, and that its response is governed by rational laws. In other words, it treats the economy as a dead object that can be modelled like a falling stone. However, a model that views the economy as a kind of super-organism would ascribe fluctuations not just to external causes but to the market itself. What makes its response unpredictable is to a large part its own inability to predict. And like a living organism, the economy represents a shifting, dynamic balance between opposing forces—positive and negative feedback loops, buyers and sellers—so models are sensitive to changes in parameterization.

In its insistence on rationality, the EMH is therefore a strange inversion of reality. Its primary aim, it appears, is not to predict the future, but to make it look like we all know what we're doing. This is dangerous for two reasons. The first is that because the EMH views the market as "normal," it gives an illusion of control, and at the same time it tends to underestimate the real risk of future financial storms. This is especially a concern when the results are incorporated in policy. As the economist Kenneth Arrow, who worked as a weather forecaster during the Second World War, put it: "Vast ills have followed a belief in certainty, whether historical inevitability, grand diplomatic designs, or extreme views on economic policy. When developing policy with wide effects for an individual or society, caution is needed because we cannot predict the consequences."[63] There is a curious disconnect between the consistent inaccuracy of the forecasts and the confidence with which politicians, banks, and business leaders regularly use them to make important decisions.

The second danger comes from the insidious idea that "the

market is always right," that it is some kind of hyper-rational being that can outwit any speculator or government regulator. This view of the economy, enshrined in the EMH, turns the market into a deity who is watched over, granted legitimacy, and explained to the rest of us by the economic priesthood. It leads to what George Soros has described as a "market fundamentalism," which is as dangerous as any other kind of fundamentalism and gives a kind of carte blanche to empathy-free corporations to do what they want, under the pretext that they are just being efficient.[64] People or countries that fail under this system have been judged by the market. But the market is no more rational than the square root of two. It is just the net effect of our own stumbling as we try to find our way in an unpredictable universe.

THREE SIBLINGS

As we've seen in these last three chapters, the scientific approaches to making predictions in the areas of weather, health, and wealth share much in common. Atmospheric and economic forecasts have always been linked, even more so when agriculture played a larger part in the economy. This was acknowledged by the Victorian scientist Stanley Jevons when he attempted to predict the business cycle by monitoring sunspots. Francis Galton used his statistical methods to study inheritance, but they have proved equally useful—and much less controversial—in economics. Today, the techniques used by scientists are essentially the same in all three areas, and all have their roots in nineteenth-century astronomy.

While celestial objects are happy to obey dictates like the law of gravity, and are amenable to modelling by equations, systems like the weather and the economy appear more anarchic. The rules they obey are local and social in nature, rather than global. As a result, in all three areas of prediction, scientists run into the same problems. The underlying system is uncomputable, so models rely on parameterizations that introduce model errors. As the model is refined, the number of unknown parameters increases. The multiple feedback loops that characterize such models also make them sensitive to even small errors in parameterization. As a result, the models are highly flexible and can be made to match past data, but accurate predictions of the future remain elusive. The models are often most useful as tools for understanding the present function of the underlying systems.

The three areas of scientific forecasting—weather, health, and wealth—are like siblings. They have the same origins, grew up together, and hung out with some of the same people. Each has its own character. Weather, the eldest, is the one the others look up to, because it is closest to the stars and knows physics. Health, the youngest, used to get in trouble, but it's flush with optimism as it prepares to come of age. (In its school yearbook, it was voted most likely to find a cure for cancer.) Wealth is the narcissist, spending its days preening in front of the mirror, in thrall to its own beauty and efficiency. In the final part of the book, we find out how these would-be clairvoyants fare as they join forces to take on the greatest challenge of all—a long-term prediction for the planet.

FUTURE

7 ► THE BIG PICTURE
HOW WEATHER, HEALTH, AND WEALTH ARE RELATED

What a chimera, then, is man! What a novelty, what a monster, what a chaos, what a subject of contradiction, what a prodigy! A judge of all things, feeble worm of the earth, depositary of the truth, cloaca of uncertainty and error, the glory and the shame of the universe!
—Blaise Pascal, *Pensées*

Past performance is no guarantee of future results.
—Mutual fund prospectus

CASE HISTORIES
The previous three chapters looked at short-term predictability in atmospheric, biological, and economic systems. Long-range forecasts, the subject of this last part, differ in that the aim is not to predict exactly what will happen at some fixed date, but to estimate major future effects. This may seem a fundamentally different task, but really the only things to have changed are the scales in time and space. Instead of predicting the local weather, averaged over an afternoon, some days in advance, long-range forecasters want to estimate the regional climate, averaged over a number of years,

some decades in advance. In medicine, a long-range forecast might address the likelihood of large-scale pandemics emerging in the global population, while in economics, it could be concerned with the scope for, and consequences of, continued growth.

These systems are, of course, not independent of each other, especially over longer time periods. Global warming, for example, is a function of carbon-dioxide emissions, which depend on economic activity. To make a prediction for our civilization and the planet, we need to consider physical, biological, and social effects as, to use Keynes's expression, an organic unity. The outcome also depends on the choices we make. Let's look, for example, at two cases where older civilizations have tangled with their environment, to mixed effect. The first is notorious, the second less so.

Case history A is the nicely named Easter Island. This small island in the Pacific is a little off the beaten track, 3,700 kilometres from the coast of South America, but it's famous among tourists and archaeologists for the amazing stone figures, the *moai,* that line the shores. When the Polynesians colonized it around 400 A.D., the island was a subtropical paradise alive with forests, birds, and animals, its seas rich with fish and dolphins that the islanders caught from canoes. Over the course of hundreds of years, a kind of small civilization grew up. The population reached around 10,000 and divided into clans and classes. Despite what we see as their remoteness, the islanders, like the citizens of Delphi, believed themselves to be at the centre of the universe. (The name of one spot translates to "navel of the world.") They honoured their ancestors by carving the giant *moai* out of volcanic rock. Transporting and erecting the incredibly heavy *moai* required a great deal of ingenuity, and large numbers of log rollers. Between this, the clearing of land for agriculture, the use of firewood for heating, and other effects, the island was by 1400 completely deforested. The birds in the forests went extinct; exposed soil blew away into the ocean; there was no wood to make canoes; streams and lakes dried up; crop yields collapsed; wars were waged over the island's remaining resources; everything went Malthusian. By the time the island was encountered by Europeans in 1722, there were only a couple of thousand people left, and they had taken up cannibalism. In the end, the islanders even turned against their stone gods, toppling and destroying them until not one was standing. Perhaps their promises, or predictions, had not come true. In 1900, after most of the remaining population had been ravaged by smallpox introduced by the Europeans, only 111 people remained.[1]

Case history B is the still smaller island of Tikopia, located just east of the Solomon Islands. Only 4.6 square kilometres in size, Tikopia was settled earlier than Easter Island. It was heading the same way until about 100 A.D., when it seems the population was converted to the benefits of orchard gardening and sustainable lifestyles. (I imagine an early, hard-core version of the Green Party.) Taboos developed that regulated both procreation and the consumption of food. Zero population growth was policy. It was enforced by the usual methods of celibacy and birth control, but also by more extreme techniques, such as abortion and infanticide (usually suffocation). Young men were sent out to sea on highly risky fishing expeditions, with the knowledge that only a few would return.[2]

The initial conditions in both cases were similar, but over time scales of centuries, the outcomes were completely different. Easter Island is a hit with tourists and a fright show for environmentalists. Many of the *moai* statues have now been restored, and they stare out from the covers of books and magazines as a kind of warning. No one will ever, ever mindlessly pollute there again. Some trees are beginning to return.[3] Tikopia still supports around a thousand souls; zero population growth is assured because the young people tend to leave.

Who would have seen it coming? The course of civilization does not run smooth; ingenuity may not translate into survival skills, and technological achievements outlast their creators. The people of Easter Island and Tikopia have achieved a kind of quasi-balance with their environment. So how will the rest of us fare as we push against the limits of our much larger but still finite island? What type of story will ours be—horror, light-hearted comedy, or difficult European art film that offers no easy answers?

It seems unlikely that a kind of global civilization model, similar to the psychohistory in Isaac Asimov's fictional Foundation Trilogy ("Q: Can you prove that this mathematics is valid? A: Only to another mathematician"[4]), can tell us the answer, given that we can't predict next week's weather. As Karl Popper asserted in 1957, "There can be no prediction of the course of human history by scientific or any other rational methods."[5] However, in 1968, the Club of Rome's thirty members, drawn from science, business, and government, had a go. In collaboration with some professors from MIT, they loaded a computer model called World3 onto a mainframe, fed in some data, and stood back to see what would happen. The results, published in *The Limits to Growth,* were Easter Island: The Sequel. "If the present growth trends in world population, food production, and resource depletion continue unchanged," the forecasters wrote, "the limits to growth on this planet will be reached sometime within the next one hundred years. The most probable result will be a rather sudden and uncontrollable decline in both population and industrial capacity."[6] The results appeared to imply that world population would peak at around 10 billion and crash to around half that in the middle of the twenty-first century.

The problem with such predictions, as the authors pointed out, is that present growth trends will not be sustained. Nothing in this world is fixed, especially not trends. The results implied that oil supplies would run out in the 1990s, but that didn't happen (in part because more oil was discovered).[7] Similarly dire warnings in the 1960s that the human population would collapse for lack of food did not come true because trends changed (population growth rates fell and food production improved).[8] So how far can we see into the future? And can scientific models help?

To answer this question, we must consider long-term predictions of weather, health, and wealth. But we must first set the stage with a brief history of these three intertwined aspects of our lives—the story so far—and then go on to discuss future projections. Climate change is a highly contentious issue, so it gets the most space. Economic predictions that extend more than a few months ahead are more futurology than science, so economic growth is here discussed primarily in the context of how it will affect, and be affected by, climate change. Finally, we take a careful peak at global pandemics.

OUR HUMAN STOCK

The easiest place to begin a prediction is with the historical charts. The earth is billions of years old, and mankind has been around for hundreds of thousands of years—unless you believe in creationism, in which case you probably believe in Armageddon as well, which kind of takes the fun out of prediction.[9] But for the rest of us, we join the story about 10,000 years ago, with the invention of agriculture in places such as the Fertile Crescent, in today's Middle East. This technological and cultural leap was in part made possible by a relatively stable climate. For 3 million years, the climate had alternated between warm and cold periods, driven by subtle oscillations in the earth's orbit.[10] Ten thousand years ago, the last ice age, known as the Younger Dryas, had just thawed out, and average temperatures had increased from about 0°C to a relatively balmy 14°C. Sun hats were back in fashion.

Agriculture spread slowly, bringing improved nutrition and fuelling a rapid increase in population (from about 5 million to around 250 million at the time of Christ). Societies, which pre-agriculture had consisted mostly of roaming bands or tribes, grew into increasingly complex and stratified civilizations, with distinct classes of priests, soldiers, rulers, and labourers. Civilizations including the Greek, Roman, and Mesopotamian empires, developed money—usually coins of precious metals like gold, silver, and bronze, stamped with the images of gods and goddesses (like Apollo).

We tend to think of environmental problems as a modern phenomenon, but any civilization will grow until it encounters a limit of some type, and often it is a natural limit. Removal of forests and over-extensive agriculture led to droughts, floods, and topsoil erosion, sometimes causing local environmental collapse. Plato was aware of the dangers and famously described deforestation in Attica: "What now remains compared with what then existed is like the skeleton of a sick man, all the fat and soft earth having wasted away, and only the bare framework of the land being left."[11] Despite such warnings, forests continued to disappear across the Mediterranean region.

Along with agriculture, massive macro-engineering projects began to leave their mark. During the Qin and Han dynasties (221 B.C. to 220 A.D.), large parts of China's forests were cut down to provide scaffolding, fuel, and housing for wooden cities and the Great Wall. The wood required during the building of the Great Pyramid of Khufu, in Egypt, came from cedar trees in Lebanon. Deforestation in many areas affected the local climate and led to permanently warmer and drier conditions—the Fertile Crescent is no longer so fertile.

The increasing size and density of cities led to rapid transmission of ideas and the development of sophisticated culture. Bigger cities also created the conditions to support and sustain epidemics. Rome in the third century A.D. had a population density similar to that of a tenement in today's Mexico City, and there was no running water or sanitation. Deadly outbreaks were frequent occurrences. The Plague of Justinian (540–42), believed to have been bubonic plague, is estimated to have killed 25 to 40 percent of Europeans. A thousand years later, the disease returned as the Black Death, with a similar effect on population. The impact of these pandemics was so large that according to one theory, it can even be measured in carbon-dioxide emissions.[12] Antarctic ice cores store air samples absorbed from the atmosphere over millennia, providing a record of atmospheric carbon dioxide; during plague years, farms were allowed to grow wild, so they absorbed carbon dioxide from the air, producing a dip in the records. Only smallpox would prove more deadly.

We didn't get the upper hand over disease until the invention, by James Watt and others, of that great cure-all, the steam engine. This kick-started the Industrial Revolution in England by making it possible to transform coal into energy. Improved economic and material conditions, along with developments in sanitation and medical techniques such as vaccination, soon resulted in significantly lower death rates. The world population expanded, reaching the billion mark in the first half of the nineteenth century. By 1930, we were up to 2 billion souls, some with their own cars, now powered by that other carbon source. Birth rates slowed in industrialized countries, but elsewhere they remained high. Population grew in a roughly exponential fashion, and this growth was supported by the Green Revolution of the 1970s, which led to more productive agriculture and increased the planet's capacity to feed the species. At the millennium, the world headcount had reached a frothy 6.1 billion. About 10 percent of the people who have ever lived are alive today. The earth has never before supported humanity on such a scale.

Economies also grew at a rapid clip. At the time of Christ, per capita income would equate to about a dollar per day (around what a person in an impoverished country can survive on today). Since then, it has increased, on average, to about fifteen dollars, with much of this increase occurring after 1700.[13] Like successful suburbanites, we have grown both larger and richer. But how is the planet doing? What would Plato or his mentor, Socrates, say now?

THE WORLD: OVER 6 BILLION CUSTOMERS SERVED

As our population and economic processes have expanded, our impact on the environment has grown. Large areas of "new" countries like Brazil and Canada are fairly pristine, but in many areas of the planet, "only the bare framework of the land [is] left." About a quarter of all ice-free land has been transformed into cropland or pasture, and much of the rest is exploited in some way for natural resources.[14] Around half the world's forests are now gone, and more have been significantly fragmented or otherwise degraded. Engineering works and other processes disrupt the earth's surface on a scale comparable to that of erosion by wind or water. All life forms transform their local environments, but our impact is multiplied by technology. Other species don't cover the land in concrete.

Our impact on the oceans is less immediately visible, since it is under water and out of sight. However, populations of large fishes such as tuna and cod have crashed by as much as 90 percent. Techniques like bottom trawling have laid waste to fragile ocean-floor ecosystems and all but depleted fishing grounds like the Georges Bank off Nova Scotia.[15] The water quality in some areas has recently improved—the Thames River, for example, is cleaner than it has been in decades—but those lakes, rivers, and streams that do not supply the wealthy with drinking water or recreation are often highly polluted. Changing rainfall patterns, combined with inadequate drainage, has led to increased flooding worldwide.[16]

In the air, fossil-fuel emissions have caused a rapid rise in atmospheric carbon dioxide. The Antarctic ice records show that the carbon dioxide level has stayed within a band of about 180 to 280 parts per million for the past few hundred thousand years—until recently. In 1958, it reached 315 ppm, and it is now about 380 ppm, and climbing.[17] Much comes, almost invisibly, from the exhaust of private vehicles; every fifty litres of gasoline contributes about 115 kilograms of the gas. Local air pollution is of course not a new phenomenon—people complained about the soot in ancient Rome. However, pollution is now a global problem, and chemically synthesized molecules, with their complex and often unknown effects on atmospheric and biological chemistry, are ubiquitous.

The waste products of civilization also affect local and global climates. Carbon dioxide is a greenhouse gas and contributes to global warming. Over the course of the twentieth century, the planet's average surface temperature rose by about 0.6°C. The 1990s were the warmest decade since record-keeping began, and the 2000s are so far following suit. Among other effects, this has led to increased forest fires across the world. Even the (normally damp) Amazon rainforest in Brazil and Venezuela is now vulnerable (as shown when a portion the size of Belgium was destroyed by fire in 1997–98).[18]

Human actions have therefore profoundly affected land, water, air, and fire. The only one of Plato's elements not to have been touched is the ether—unless we count the electronic signals beamed around the world by satellite. And we have an even greater impact on life. While we as a species are doing well, our effect on creatures of the land, oceans, and skies has often been disastrous. We prosper as other species crash.[19] Every extinction knocks another set of genes out of the world gene pool and affects the robustness of the global ecosystem.

Even this low-resolution history of the world is enough to show that we are living in unusual times: what we take as normal isn't that normal. The climate is always prone to change, but over the past 10,000 years, humanity has enjoyed a warm and relatively stable period that has been suitable for the development of agriculture. We seized that window of opportunity, and now we have put almost all suitable land to the purpose of feeding ourselves. The length and quality of life in many countries has as a result increased enormously, but at the expense of degrading the existing air, water, and soil. And while our stone-age ancestors could react to land loss or shifting climates by moving, we don't have that flexibility; there's little slack in the system. As people in poor nations cluster in vulnerable areas, they are increasingly susceptible to natural disasters, such as the storm that hit Venezuela in 1999.

Many people in industrialized countries have grown up with a greatly reduced fear of infectious disease. Antibiotics and improved sanitation were a great success story of modern science, and they have vastly reduced death rates in many countries. The last really serious influenza pandemic to hit the rich world, in 1918, killed at least 20 million and made many more ill. Bubonic plague has been all but eradicated; smallpox exists only in the lab.[20] Of course, the gains won were not permanent or global. The number of people infected with HIV/AIDS worldwide is about 40 million. New diseases such as SARS continue to emerge, and overuse of antibiotics has led to resistant strains of old diseases like tuberculosis.

Perhaps the most unusual thing about recent history, though, has been the extraordinary rate of economic growth. Industrialized countries like the United States commonly try to achieve annual growth rates in gross domestic product of around 3 or 4 percent. If the *homme moyen* had pulled in a dollar per day 2,000 years ago and a growth rate of only 1 percent over inflation was maintained since, he would today be enjoying a healthy pay packet of $439 million per day.[21] Since not everyone plays NBA basketball, this is clearly impossible. Economic growth is the relatively recent by-product of the Industrial Revolution. In rich countries, we have until now had the best of all possible worlds: a good climate, excellent health, and an explosive economy. In bread-making terms, we are like a yeast colony in a dough: carefully nourished with all the requirements for life, covered with a towel, left undisturbed in a warm place, and growing exponentially. So how long can this go on?

LIVING IN A BUBBLE

According to chart-following optimists, the answer is forever. After all, many worriers have cried wolf in the past, and their dire predictions never came true. Malthus thought the world would disintegrate into famine, plagues, and wars, but while these have certainly happened, none has come close to stopping the growth of the total world population. We have the wind in our sails, momentum behind us, and nothing can hold us back. Global warming and overpopulation are just the fevered inventions of a neurotic society that is living in the safest period of human history and is egotistical enough to believe its challenges are unique.[22]

One such optimist is Michael Crichton, the author of thrillers such as *Jurassic Park* and the creator of the hit television hospital drama "E.R." For his 2004 thriller, *State of Fear,* he spent three years researching climate change and the environment, and he came in on the side of the skeptics. In an appendix to that novel, which created a lot of controversy and was cited in the U.S. Senate as a useful contribution to the global-warming debate, he argued that climate predictions are hopelessly unreliable, and that improved technologies will mean that we never run short of resources. "For anyone to believe in impending resource scarcity, after 200 years of false alarms, is kind of weird," he wrote. "I don't know whether such a belief today is best ascribed to ignorance of history, sclerotic dogmatism, unhealthy love of Malthus, or simple pigheadedness."[23]

Indeed, we have been told many times that we are on the verge of running out of oil or water or some other resource. The United States Bureau of Mines predicted in 1914 that American oil would run dry within ten years. Similarly alarmist predictions were made by the Department of the Interior in 1939 and 1951 (thirteen years of U.S. oil left), and by the Club of Rome in 1972 (not much of anything left). All, apparently, were proved wrong. As Adam Smith knew, scarcity means that the price goes up, so either new sources are found, which is what happened with oil (the Alberta tar sands are big), or an alternative is developed.[24] In their book *The Bottomless Well,* Peter Huber and Mark Mills argue that energy use is positively virtuous, because "energy begets more energy.... The more energy we seize and use, the more adept we become at finding and seizing more."[25] According to this theory, when we approach a stop sign, we should step a little harder on the gas.

Some scientists have always believed that science itself will learn to control the progress of the human race and steer us to safety. In *A Masculine Birth of Time,* Sir Francis Bacon wrote of science bringing about "a blessed race of Heroes and Supermen." As the psychologist B. F. Skinner wrote in 1973, "What we need is a technology of behaviour. We could solve our problems quickly enough if we could adjust the growth of the world's population as precisely as we adjust the course of a spaceship."[26] And if that doesn't work, and we run out of room, we could always try a real spaceship and find another planet to colonize.

Optimism in the power of science and progress is heartwarming, but those pigheaded people who believe in fundamental analysis tend to take a more jaundiced, seasoned view. They will point out that recent successes took place in a very favourable environment, which may not continue; that these successes have been mostly limited to rich countries and have done little for a substantial proportion of the population; that excessive expansion often results in overshoot of fundamental limits (i.e., a bubble); and that when bubbles burst, they do so with a bang. In other words, all the boosterism for human ingenuity and technology is about as meaningful as that heard for the latest tech firm at the turn of the millennium.

These fundamental analysts will cast a questioning eye over S-shaped curves such as that shown in figure 7.1 (see page 282). Most current prognosticators have the world population shrinking slightly in rich countries and growing progressively more slowly in developing countries, until it gently coasts to an equilibrium around 2150. This is based on the observation that as developing countries get richer, the death rate decreases and population goes up. After a while, the birth rate also decreases (rich people have fewer children, for some reason), so growth slows or levels out.

Because the time scale for human reproduction is slow, population projections can be relatively accurate thirty or so years in advance—but only if nothing unusual happens (such as the baby boom, which forecasters failed to predict).[27] Like economic models, demographic models tend to extrapolate the past and do not capture major turning points, like the one that is supposedly coming up. A fundamental analyst will therefore ask if this estimate of future growth, based on current trends, can be reconciled with the actual number of people the earth can support over the long term (its carrying capacity). If not, then we are living in a bubble. Rather than taper off at a sustainable limit, the population will overshoot and then crash—as has happened time and again on a more local scale with previous civilizations.[28]

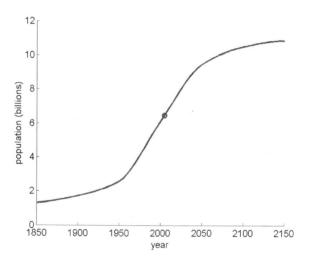

FIGURE 7.1. Total estimated world population.[29]

The carrying capacity depends on, for example, whether everyone will be living at current First World standards. Its calculation implicitly assumes all kinds of things about economic and technological growth and the resilience of nature. (Even the definition of "sustainable" is nebulous. I will take it to refer, like homeostasis, to maintaining a kind of dynamic metabolic balance with the environment. Sometimes it is easier to define in the negative—Easter Island, not sustainable.) For everyone to enjoy a Western lifestyle, one estimate pegs the carrying capacity at around the 1930 population of 2 billion.[30] This might seem low, but it accounts for the fact that rich people have far more impact on the planet than poor people. The environmentalist World Wildlife Fund estimates that by 2050, we will need almost three earths to support ourselves in the style to which we have become accustomed.[31] Others believe that new technologies will allow us to support a much greater density, even higher than today's. What is evident is that we are currently degrading the biological and physical systems that support life. In the words of the United Nations' 2005 Millennium Ecosystem Assessment, we are "living beyond our means."[32]

There is a chance that our technology will improve and growth rates will slow and reverse, steering the population back to a sustainable level; but there is also the possibility of a dramatic collapse. Maybe the Club of Rome was right. If so, then environmental skeptics are not skeptics at all—they are boosters who make naïve extrapolations based on recent performance. The true skeptics are people like Sir Martin Rees, Astronomer Royal, who pegs our civilization's chances of survival to the end of the century at only 50 percent.[33] (His predecessor in that position, John Dee, was also interested in predictions of the apocalypse, but his were based on the whispering of angels.[34])

One study on opinions about climate change revealed a striking cultural divide between natural and social scientists.[35] The latter, which includes conventional economists, tend to believe that the impact of even severe global warming will be low, and that we can invent replacements for "climatic services." The former, which includes climate scientists, are far more pessimistic, and estimate the economic damage of global warming to be twenty to thirty times greater, with much of it assigned to non-market categories like nature or quality of life. Perhaps economists have more at stake in the existing economy.

So if we can't even agree on the scale of the problem, what can computer models tell us about the future? Is it possible to make accurate predictions for the planet, or will the question be bounced back and forth between optimists and pessimists until time tells us the answer? And what can we say about the implications for our weather, health, and wealth?

WEATHER: GETTING WARM

While our impact on the planet is combinatorial, one issue that has received much recent attention and controversy is climate change caused by the accumulation of greenhouse gases in the atmosphere. We still fear the weather gods, and the more extreme possibilities—all-destroying hurricanes, killer droughts, apocalyptic floods—evoke images of almost biblical wrath. Sudden changes in climate have always been a risk—a thought here for the pre-Inca Moche of the Andes, or the Maya of Central America, whose water supplies were cut off by shifts in weather patterns—but we may become the first civilization to be affected by our own actions on a global scale. The scientist and environmentalist David Suzuki said, "Climate change is one of the greatest challenges humanity will face this century."[36] Meteorologists, meanwhile, see global warming as a tractable and interesting theoretical problem that can be addressed by the use of large mathematical models. But is it possible to predict the climate if we can't predict next week's weather?

The basic physics behind global warming was pointed out by the Swedish chemist Svante Arrhenius in 1896.[37] The earth is warmed by energy from the sun, which arrives as a broad spectrum of electromagnetic radiation, including visible sunlight. Bodies at the earth's temperature in turn emit long-wavelength, infrared radiation (which is how soldiers can detect people in the dark using night-vision goggles). Instead of radiating out into space, this energy is mostly absorbed by the atmosphere. The degree of absorption depends strongly on the exact concentrations of certain gases. These are known as greenhouse gases because they have the same effect as the glass in a greenhouse does: they let the light in, but don't let the heat out. This is not a subtle effect. Without greenhouse gases in the atmosphere, the average temperature would be $-18°C$ instead of $+14°C$. We have all been raised in a hothouse.

The greenhouse gases, which include water vapour, carbon dioxide (CO_2), and methane, are therefore vital to our survival. However, you can have too much of a good thing. Increasing CO_2 from its pre-industrial levels of about 280 ppm to 380 ppm is a substantial relative change. And even if we were to freeze CO_2 emissions at current levels, its slow rate of decay means that the total amount will still continue to grow well into the future. Furthermore, because of the slow response of the ocean/atmosphere system, the effects of high CO_2 levels will be with us for centuries, and may even be irreversible. A common property of non-linear systems is hysteresis: once a change has been made, it is difficult or impossible to undo.

In 1979, the meteorologist Jule Charney organized a meeting at Cape Cod to investigate what would happen if CO_2 levels were double the pre-industrial level. (This may occur sometime during the present century, with climate effects tending to lag behind as the system adapts; though as discussed below, the exact rate of CO_2 production depends on a wide range of social and economic factors.) At that time, there were only two American research groups actively involved in climate modelling: Syukuro Manabe's at the Geophysical Fluid Dynamics Laboratory (GFDL) at Princeton, and James Hansen's at NASA's Goddard Institute. Manabe had been involved with climate models since 1963. (These are usually simplified versions of GCMs, with some of the details stripped out so they can be run over periods of hundreds or even thousands of years.) On the first day of the meeting, he told the assembled group that according to the model, a doubling in CO_2 would lead to a rise of 2°C—not too bad. The next day, though, Hansen presented the results of his model: it indicated a much larger rise of 4°C, a factor of two difference. Since the aim was only to get a rough idea of the possible magnitude, Charney chose 0.5°C as the margin of error on both calculations, which left a range of 1.5°C to 4.5°C (Arrhenius had estimated 5°C back in 1896). The lower limit was not out of line with natural variations that have occurred in recent centuries, but the upper was a real ice-cap melter. In words, the estimate meant that either nothing much could happen or we could have a serious problem.[38]

Not everyone at the meeting was happy with the agreed range, and some described it as hand waving.[39] However, it did seem to indicate that CO_2 could have a considerable effect on the planet, and it helped mobilize scientific interest. The Intergovernmental Panel on Climate Change (IPCC) was established to report on the matter. At its second meeting in 1995, there were thirteen climate models to choose from; by the time of its 2001 meeting, climate modelling had grown into a major research area for groups all over the world, including the Max Planck Institute for Meteorology in Germany, the Hadley Centre in England, and the Lawrence Livermore National Laboratory in California. In a remarkable display of consistency, though, the estimate of potential warming, obtained by consensus among experts, remained little changed from what is sometimes called the "canonical" range of 1.5°C to 4.5°C, with an average of about 3°C.[40] (The panel did not assign probabilities to the different outcomes, because there was no sensible way to determine them. One of the scientists observed, "The range is nothing to do with probability—it is not a normal distribution or a skewed distribution. Who knows what it is?"[41])

Some more recent sensitivity studies indicate a rise beyond that canonical range. As the climatologist Stephen Schneider noted, "Despite the relative stability of the 1.5 to 4.5°C climate sensitivity estimate that has appeared in the IPCC's climate assessments for two decades now, more research has actually increased uncertainties!"[42] This degree of uncertainty reflects the challenge of modelling the climate system. Just as a biological system incorporates complex regulatory loops that make it difficult to model, the climate system is made up of feedback loops that elude simple parameterizations.

GLOBAL COOLING

Can the future climate be predicted by looking at past climates? An alternative to using mathematical models is to adopt a data-driven approach, and study how greenhouse gas concentrations and climate have varied together with the ebb and flow of past ice ages—a field known as paleoclimatology. Sources of data include ice cores (which trap gases), petrified tree rings, and geological features such as sedimentary rocks.

There are two difficulties with this approach. First, the data contains a high degree of uncertainty. Scientists argue over how to interpret satellite pictures, let alone tree rings. Second, there is no exact analog in the past for the current situation—dinosaurs didn't drive cars. Estimates of climate change based on paleoclimatology are therefore highly uncertain. As might be expected, they have a tendency to fall in the "canonical" range. Some scientists, though, point to extreme events in the past as proof that the climate system is more sensitive than models allow.

One such extreme event occurred during the Younger Dryas period. This was named after a small, cold-loving plant—*Dryas octopetala*—which pollen records show suddenly began to flourish across much of Europe about 12,700 years ago. The cause, it seems, was a sudden and precipitous drop in temperature, which plunged the warming earth back into ice-age conditions for over a thousand years. Some believe that it could have been triggered by an abrupt reversal of the ocean circulation pattern that drives the Gulf Stream, and that melting Arctic ice could initiate a similar event in the future. Global warming may mean that some areas, such as the U.K., become much cooler.

IN THE LOOP

As mentioned in Chapter 4, one of the key processes involved in determining our weather is the formation and dissipation of clouds. They play an equally important role in the global climate. If the planet warms, more water evaporates into the atmosphere. Since water vapour is a greenhouse gas, this can trap heat and lead to further warming—positive feedback. If the water vapour condenses into clouds, though, these create shadow and lead to cooling—negative feedback. Except at night, when cloud cover prevents heat from escaping—positive feedback.

Water in the form of snow and ice also affects the planet's albedo, which is its ability to reflect light. White surfaces have high albedo, and dark surfaces have low albedo (the word comes from the Latin for whiteness). Most of the earth is blue, green, or brown, colours that are somewhere in the middle; but snow, being white, reflects about 80 percent of the sun's energy. It therefore acts to keep the planet cool. Antarctica is white all year-round, but in the Arctic, snow and ice cover is seasonal and sensitive to changes in climate. If temperatures increase, the cover reduces and the albedo goes down. The ocean, deprived of its ice layer, also releases more water vapour to the atmosphere, which accentuates the greenhouse effect. Together these cause further warming, in a positive feedback loop known as the Arctic amplification. Small differences in ice albedos have a large effect on model results.[43]

In fact, climate models must contend with feedback loops that are as complicated as those in any biological system—and usually have a biological component. Another example is the carbon cycle. Carbon is the most important building block for life. Its atomic structure gives it a unique ability to combine with other elements to form complex molecules such as DNA. Like the blood in our bodies or money in the economy, it is constantly being cycled around the biosphere. Plant life uses carbon dioxide for photosynthesis, combining it with water to create sugar and oxygen. The latter is breathed back out to the atmosphere.[44] Algae on the ocean surface and other sea organisms similarly absorb huge amounts of carbon for photosynthesis. (Algae are also believed to be cloud-makers—they emit sulphur gases, which oxidize to form minute particles that encourage the formation of clouds.)

We too are now a major part of the carbon cycle: fossil-fuel emissions and cement manufacture release about 5.5 billion tons of carbon per year, while deforestation and other land use contribute another 1.5 billion tons.[45] Furthermore, global warming can affect carbon levels in a positive feedback loop, by increasing the number of forest fires and reducing the amount of CO_2 held in ocean water. Large amounts of carbon, as much as 450 billion metric tons, are also stored in the frozen tundra and boreal forests of the North.[46] We have to hope that they stay frozen, since their release could lead to a runaway warming.

The climate system therefore consists of a nested series of non-linear feedback loops that are in a kind of dynamic balance. Unlike short-term weather models, climate models do not suffer from sensitivity to initial condition, since the aim is to predict the average weather far in the future (which should be insensitive to the exact starting point). This is especially so if the model is run for a long period under fixed conditions, in which case it settles onto its own attractor. Error in climate models is therefore primarily a result of model error.[47] If GCMs were perfect, so that "predictability limitations [were] not an artifact of the numerical model,"[48] then climate prediction would be easy. But this is far from being the case. Weather and climate predictions are directly linked, and to believe that models can fail at the former but succeed at the latter is nothing but wishful thinking—especially when the most basic properties of the ocean/atmosphere system prove so difficult to model.[49]

WATER WORLD

One substance that consistently slips through the grasping fingers of weather and climate modellers like water is water. Because water is so ubiquitous, we don't often reflect on its properties; but it is perhaps the most mysterious, upside-down, shape-shifting, and life-giving substance on earth. If carbon represents the earth's yang, fixing and organizing other substances into its geometry, then water is its yin. The two are as different as water and oil (which is 85 percent carbon).

As Antoine Lavoisier demonstrated, a water molecule consists of a single molecule of oxygen (O) and two molecules of hydrogen (H). The frost on the ground, the mist in the air, the water in our bodies—all have this simple chemical structure. The molecule is highly polarized: one side has a positive charge, the other a negative. Since opposites attract, the positively charged side is drawn to the negatively charged side of nearby molecules, in a process known as hydrogen bonding. Collections of water molecules form a complex interacting group, each molecule aligning itself with the others in a vibrant dance, switching partners billions of times per second. This gives the substance a number of unique properties. For example, the solid form is less dense than the liquid form, so ice creates a protective layer that insulates the water below. From the level of the single cell to the entire globe, water is essential for life. About 70 percent of a typical cell is water, about 70 percent of our body weight is water, and about 70 percent of the earth is covered in ocean.

In the climate system, water appears in far more guises than is suggested by the simple categories of solid, liquid, and gas. In its solid form, it might be ice or snow or different types of crystal in the atmosphere. As a liquid, it can appear as rain, fresh water in rivers and streams, or brine in the sea. Water vapour can evaporate from the oceans or land or be absorbed by plants, and it can condense into every different type of cloud that you see in the sky as a mix of vapour, liquid, and crystals. And water constantly changes its form: a single molecule might start off in the ocean, evaporate into the atmosphere, join a passing cloud, fall to the ground as snow or rain, enter a river, and flow back into the ocean, all within days. As Leonardo da Vinci wrote, "The waters circulate with constant motion from the utmost depths of the sea to the highest summits of the mountains, not obeying the nature of heavy matter; and in this case it acts as does the blood of animals which is always moving from the sea of the heart and flows to the top of their heads. . . . These waters traverse the body of the earth with infinite ramifications."[50] Each ramification has a different effect on the climate. Water is therefore fluid, non-linear, constantly changing its appearance, and generally un-Pythagorean.

CLIMATE ATTRACTOR

Because of the importance of water in the global climate system, model predictions depend on exactly how its behaviour is represented. GCMs generally agree that as global warming continues, more moisture will evaporate and form clouds, therefore affecting the planet's albedo; however, the exact change in albedo depends critically on the type of clouds.[51] Climate modellers may in the future be able to hammer out a consensus on these and other difficult points, but there is no guarantee that the consensus will be right, because no model can capture the full scope of the climate system.

A typical GCM used for climate studies might divide the atmosphere into a three-dimensional grid with cells about the size of a small country like Belgium. For a particular property, such as albedo or cloudiness, they assign a single number or simple distribution to everything within a cell. Like fractals, though, natural surfaces reveal greater amounts of detail the more you zoom in. There is never a point where more resolution doesn't reveal a finer degree of structure.[52] Since GCMs can offer only rough parameterizations of the physics of clouds—or the growth of trees, or changes in grasslands, or the deformation of ice under pressure, or the exact way water evaporates from exposed ground, or a whole host of other things—they are not true physics-based models. They cannot be demonstrated from fundamental laws. Like models of the economy, they are really a collection of approximate equations that have been combined and balanced to give reasonable results.

Many climate models, if run forward for thousands of years, would not give a realistic-looking climate at all without the use of so-called flux adjustments. Left to their own devices, models will cheerfully boil away all the water in the oceans or cover the world in ice, even with pre-industrial levels of CO_2. (The fact that the earth doesn't do this says something interesting about its regulatory networks, as we'll see in the next chapter.) The only way to bring them back to something reasonable is to somehow fudge the heat balance—altering, for example, the transfer of heat between ocean and atmosphere.[53]

The models are also strongly affected by changes in parameters. In one recent Oxford University–led experiment, the largest of its kind, several key parameters controlling the representation of clouds and precipitation in a GCM were set to alternative values "considered plausible by experts." The effect was to explode the range of predictions, pushing it as high as 11.5°C (after omitting some simulations that became unstable, or even showed cooling).[54] The climate system consists of an intricate balance of opposing feedback loops—what Heraclitus described as a "harmony of opposite tensions"—and small changes in their representation can have large effects. This sensitivity does not imply that the climate system itself is unstable. All that can be concluded is that the models are sensitive to parameterization.[55] The average rise was close to 3.4°C, but this only reflects the fact that perturbations were made around a model that happened to show that amount of warming.

Even such experiments do not reveal the true uncertainty in the calculation, because the very structure of the GCM equations fails to capture the underlying dynamics. Parameters are properties or constructs of the model, not the system. As one paper put it, "As soon as we begin to consider structural uncertainty, or uncertainty in parameters for which no prior distribution is available . . . tidy formalism breaks down. Unfortunately, the most important sources of model error in weather and climate forecasting are of precisely this pathological nature."[56] Since the errors are "pathological" (i.e., not expressible by equations), it is as difficult to estimate the correct parameter range, or forecast uncertainty, as it is to predict the climate itself, and for exactly the same reasons. We don't have the equations. In an uncomputable system, they don't exist.

As the pioneer climate scientist Manabe put it, "Uncertainty keeps increasing with the more research money they put in. . . . It hasn't gotten any better than when I started forty years ago."[57] Perhaps this is why, despite huge advances in computer speed, earth-observation techniques, and climate research, the IPCC and most climate scientists still quote the canonical band for carbon doubling of 1.5°C to 4.5°C.[58] This range gives an image of stability, and it helps, as the Oxford scientist Steve Rayner described it, to "domesticate climate change as a seemingly manageable problem for both science and policy." The estimates for global warming seem to say as much about the dynamics of science as they do about the dynamics of climate.[59]

REASONS TO BE VALID

Below are some common arguments for the validity of climate models—and the reasons why they are not valid.

"The models are derived from basic laws of physics." There are no laws for the formation and dissipation of clouds or many other processes.

"The models can reproduce the current climate." Versions with different parameters can adequately reproduce the current or recent climate, while giving a very different response to doubled carbon dioxide. It is always possible to tune models to fit past data; it's much harder to predict the future.

"The models can simulate past climates." Same problem. Also, there is a great deal of uncertainty in estimating climates in the distant past.

"The butterfly effect does not apply to long-term climate forecasts." This isn't very relevant, since chaos is rarely the most important factor in the modelling of physical systems.

"It is easier to predict climate statistics than next week's weather." If the climate is completely stable, this is obviously true. But if the climate is changing, it should be no easier to predict than the short-term weather. This is reflected in the huge uncertainty in climate change estimates.

"The models don't work now, but they will in ten to fifteen years." That's what was said ten to fifteen years ago, but uncertainty has only increased. We may have a better idea by then of what global warming entails, but it probably won't be because of improved models.

"Different model versions cover everything from no warming to more than 11°C, so one must be right, at least for the global average." True, but it doesn't mean the models are right—just that they are very flexible.

> "Criticisms are politically motivated." Avoiding criticism by brushing it off as political is political.
>
> Of course, the question of model validity is distinct from that of global warming. As in most areas of life, it is possible to believe there is a problem without claiming to be able to predict the future or have an accurate model.

THE BUTTERFLY EFFECT

This is not to say that it is impossible to make any kind of sense out of the global climate system, or to make any predictions. To take an example from a different context, suppose that a dietician monitors a child who has taken to eating a number of candy bars a day. It would be impossible to compute or predict the exact effect: in some children the candy will speed their metabolism so they burn off the energy (negative feedback), while in others it will slow them down and trigger much larger weight gains (positive feedback). However, the dietician can make an educated guess, and she would certainly expect the child to either gain weight or stay at the same weight, but not to lose weight.[60]

If measurements of the child's weight over a few weeks showed that it was indeed increasing, while factors such as the amount of physical exercise remained more or less the same, then the dietician would have some confidence in saying that the candy was a likely cause. And if blood tests showed elevated levels of glucose, then immediate action would be called for. Glucose is present in our blood at a low concentration and is closely regulated by the body. If it doubles to twice its normal level, diabetes is suspected.

Similarly, scientists since Arrhenius have known that increased greenhouse gases will have a warming effect on the climate (which is why they are called greenhouse gases and not refrigerator gases). Observations of the ocean's heat content, the worldwide retreat of glaciers, the melting of sea ice and permafrost in northern regions, and the rise in mean global temperatures all show that the planet is heating up in a manner consistent with an enhanced greenhouse effect.[61] This in itself does not prove that increased carbon emissions are the cause, since the climate naturally fluctuates of its own accord, but it certainly makes it more plausible. Predicting the exact reaction, however, is not possible: the planet may limit the effect of greenhouse gases through negative feedback, or it may pass a threshold that leads to runaway warming or triggers a sudden shift to a different climate state. And tinkering with complex feedback loops—for example, by destroying tropical rainforests—may make the climate behave in a more erratic fashion, like a mutant strain of yeast without its control mechanisms.

One indicator of changes in the climate is the migration of species such as butterflies. The Edith's Checkerspot butterfly, which lives along the Pacific coast and has been documented since 1860, has already begun moving north and to higher elevations. Its range now extends much deeper into British Columbia and Alberta. At the same time, colonies are disappearing from California and northern Mexico.[62] If global warming continues, many species will be trying to do the same thing. Tropical diseases such as malaria may be carried right out of the tropics by their mosquito hosts. But some animals, like polar bears, will have no farther north to go. And of course, people will need to migrate if their situation becomes untenable.[63]

Such effects can in principle be assessed by ecosystem models (see Appendix III for a conceptual example). These attempt to capture the interactions between various species and their environment, and can be coupled with the output of GCMs. However, the uncertainty in the climate-change predictions—which are even worse at forecasting local effects than global trends[64]—is magnified by the great uncertainty in the ecosystem models. The results are therefore speculative, but they may help identify species and ecosystems at risk. For many species, the main problem is less global warming than the fact that *we* are eating large numbers of them (fish, for example).[65]

WEALTH: HOT ECONOMY

Global warming is directly linked to one species in particular—our own. The IPCC projections have all assumed a doubling of carbon-dioxide levels. When exactly this will happen depends on emission rates, which in turn depend on the course of the global economy. This is subject to even more vagaries than the global climate, and it's even harder to predict.

To give one example: policy-makers would like to know how the stock market will grow over, say, the next seventy-five years so they can fund retirement programs. Economists can produce estimates based on a combination of numerical equations and historical analogies, but long-term predictions are no easier than short-term ones. (There is just a better chance that no one will remember your forecast.) Estimates of long-term U.S. stock-market returns typically vary from around 4.5 percent to 7 percent, which compounded over seventy-five years represents more than a factor of five difference.[66]

Perhaps for this reason, the IPCC decided, for its climate calculations, to study an array of different scenarios or storylines. These ranged from A1F1—whose "major underlying themes are convergence among regions, capacity building and increased cultural and social interactions, with a substantial reduction in regional difference in per capita income"—to B2, where "the emphasis is on local solutions to economic, social and environmental sustainability."[67] It is impossible to know which scenario is more probable, so all were assumed to be equally likely. The estimated uncertainty owing to different economic scenarios, for a fixed model, turns out to be about the same as the uncertainty owing to different models. This is despite the fact that the latter depends on things like the parameterization of clouds, while the former depends on the parameterization of Chinese consumers. When both sources of uncertainty are combined, the predicted range of climate warming for the year 2100 increases to 1.5°C to 5.8°C.[68]

The chosen economic scenarios add to the possible controversy. In 2005, one area of strong debate was whether the scenarios should rely on market-based exchange rates or purchasing-power parity.[69] Some economists believe the latter results in unrealistic projections for economic growth, and therefore emissions growth. This is a fair point, and it highlights the need to treat predictions in a holistic manner. However, the fact that climate-change predictions are highly dependent on such effects is another sign of their sensitivity, and it means that results depend heavily on the biases of those doing the calculations.

The same issue affects calculations of the effects of global warming on the economy. We often read in newspapers about how global warming could benefit certain regions. When Arrhenius first estimated the effects of global warming, he thought it would be a good thing because he lived in Sweden. Winters in my hometown of Edmonton might get shorter and milder, which is something we could probably adapt to. Before signing the Kyoto Protocol, Russia's Vladimir Putin joked that a little warming might not be a bad thing for his part of the world. The U.S. government's complacency about climate change, evident in the Bush administration's unwillingness to sign the Kyoto treaty, may be based on a similar calculation made by the Pentagon. Their scientists predicted that extreme climate change would bring about a period of war, conflict, and instability, but they added that "with diverse growing climates, wealth, technology and abundant resources, the United States could likely survive shortened growing cycles and harsh weather conditions without catastrophic losses. . . . Even in this continuous state of emergency the U.S. will be positioned well compared to others."[70] China and India would be more vulnerable to agricultural losses and population displacements, while Europe would have to cope with floods of refugees from North Africa and elsewhere.

From the Pentagon, global warming begins to sound like Von Neumann's vision of the weather as a weapon of war. While it is obviously true that a change of any sort will affect some more than others, calculating the net effect on society is not easy. One study attempted to predict the total cost of global warming using the Regional Integrated Climate-Economy (RICE) model, an offshoot of the Dynamic Integrated Climate-Economy (DICE) model.[71] Assuming a business-as-usual scenario, in which no action is taken to prevent global warming, the cost is $4,820 billion. But if humanity takes the optimal course of action, the total cost is found to be $4,575 billion, a net saving of only 5 percent. It has been claimed that this estimate is reliable because it agrees well with other similar models.[72]

The problem here is that if GCMs cannot predict the climate and economic models cannot predict the next recession, then the uncertainties only grow when the two are combined, and the results are certainly not reliable to within a few percent. When agreement does exist between different models, it says more about the self-regulating group psychology of the modelling community than it does about global warming and the economy. It is an illustration of why ensemble forecasts can be highly misleading: an ensemble of wrong models does not make a right model, and the spread between the results is not an accurate measure of uncertainty.

When the economist Kenneth Arrow was working as an air force weather forecaster during the Second World War, he and his colleagues found that their long-range predictions were no better than random. They informed the boss but were told, "The commanding general is well aware that the forecasts are no good. However, he needs them for planning purposes."[73] We can't exactly predict how the climate will change. In fact (here I agree with the random walk theory), to estimate the economic effects or precise causes you may as well toss the DICE. Projections may be useful for policy-makers, as a device to provoke ideas and aid thinking about the future, but they should not be taken literally. As Keynes once said of unforeseeable political events, "there is no scientific basis on which to form any calculable probability whatever. We simply do not know!"

Given the potential downside risk of global warming, perhaps the best approach is that of Warren Buffet, whose insurance companies, General Re and National Indemnity, are exposed to any increased risk of hurricane damage. As he told shareholders in 2005, it is unknown whether global warming will lead to more storms like Katrina, but "recent experience is worrisome. . . . Our ignorance means we must follow the course prescribed by Pascal in his famous wager about the existence of God." Even if we're not convinced about climate change, it would be prudent to pretend we are.

REASONS TO BE SKEPTICAL

Here are some arguments, heard from skeptics, for why predictions of climate change and environmental collapse are wrong—and some equally skeptical replies.

"Mathematical models of climate change are hopelessly unreliable." Models of housing bubbles and disease epidemics are also unreliable, but these things still happen.

"It is ridiculous to believe that our puny species can affect the balance of an entire planet." I wonder if the smallpox virus was plagued by similar doubts as it ran rampant through the New World. "Is it possible that I, a simple virus, can destroy a human being, let alone entire *societies*?"

"It is egotistical to think that we live in a special time, with unique challenges." This argument might have been raised by Easter Islanders before they cut down the last tree. *Easter Islander 1:* "This is the last tree. If we cut it down, there will be no more." *Easter Islander 2:* "Gee, what an ego." (Fells last tree.)

"Human ingenuity will solve the problem." Feel free to start any time.

"Environmental scare stories have consistently turned out to be wrong." Some have—but how do we know this isn't the moment in the horror film when it turns out the geek was right?

"Models have shown that the economic benefits of a warmer planet will balance the harm caused." We're skeptical about climate models, but not about economic models?

"When I put my head out the window, the air is fresh, the trees are green, there is no sign of imminent environmental collapse." I'm guessing there may also not be much visible proof of famine and extreme poverty, but apparently they do exist.

"CO_2 is present only in trace quantities in the atmosphere, and it's a carbon source for plants. What can happen if it doubles?" The carbon source known as glucose is present only in trace quantities in our blood. Double it, and you have diabetes. Triple it, and you may lose consciousness. No one can manage the environment, but just as we do with our own bodies, we can monitor health, practice moderation, and limit exposure to toxins.

HEALTH: NEXT YEAR'S DISEASE

While climate change may turn out to be a major threat to our societies and economies, an equally serious concern is our much older enemy, disease. Figure 7.1 (see page 282) shows a smooth increase in population from 1850, but a look at the time preceding that would show more of a roller-coaster ride. The population in Britain in 1348 was about 3.7 million; it dropped to 2.1 million in 1430 as a result of the Black Death and didn't recover to previous levels until 1603.[74] It is only since the Industrial Revolution that the human stock has been on a steady upwards trend. In the rich world, we now live longer and healthier lives than at any time in history. Just as climate change affects everyone on the planet, though, no population will be immune to global pandemics. This is especially true in our highly connected modern societies and economies. Diseases can be transported around the world in mere days, before any health organization has had time to react. And hastily imposed quarantines and other measures could bring global trade to a halt and devastate the world economy.[75] The next major storm might be biological, not atmospheric or financial. In this section, we turn our attention from the large scale to the very small.

A good way to learn about prediction is to study how our own bodies resist disease. Our immune system has developed over millennia to identify and counteract pathogens such as bacteria and viruses. The latter are not autonomous living beings but packaged strings of genetic information (DNA or RNA) that invade the cells of other organisms. Once inside, they hijack the cell's machinery to reproduce themselves. This often kills the cell, at which point the virus is released to find new cells to attack.

Kepler believed, at one time, that the universe was structured after the Platonic solids. While he later changed his mind on that score, he would have been interested to find that a broad class of viruses, including those for polio and the common cold, contain their genetic information in an icosahedral container known as a capsid. At each vertex are cell-surface receptors, like microscopic spikes, which attach themselves to the cell to be invaded. The influenza virus, named after the Latin word for influence (because epidemics were thought to be influenced by the stars), needs a set of only eight genes to construct itself.

The immune system's task is complicated by the fact that bacteria and viruses evolve at a much faster rate than humans. Micro-organisms are on the fast track of evolution, while our immune systems lumber along, always one step behind. Bacteria cells can reproduce in about twenty minutes (as opposed to twenty years for humans), and they happily swap portions of DNA, including those that grant immunity to antibiotics. Viruses too are unstable, so this year's flu bug may be quite different from the one that was causing problems last year. Like hit singles on the radio, they have a finite span before their novelty wears off.

The immune system must also be able to distinguish between micro-organisms normally resident in the body and foreign invaders. About 10 percent of our body's dry weight consists of bacteria (in the gut, skin, and elsewhere), and on the whole, they provide very useful functions. Taking a broad-spectrum antibiotic can affect digestion by depleting the bacteria that form an essential part of the digestive system. To recognize foreign bacteria that may be harmful—those terrorist cells—the immune system must predict what such an intruder would look like.

In humans, the first line of defence is the innate immune system. This includes the white blood cells, which are lined with receptors that recognize certain features of microbial invaders, such as components of the cell wall. When an unwelcome intruder is recognized, the white blood cells engulf and annihilate it. They also trigger the production of cytokines, which produce inflammatory responses such as fever. This system can occasionally cause problems of its own, especially when it overreacts to its own predictions; this is what happens with allergies, where the body seems to be launching some massive shock-and-awe attack on a relatively harmless substance like pollen.

While the innate immune system is always ready with a quick response, the acquired immune system takes the slow, thoughtful approach. It doesn't just eliminate intruders—it tries to get to know them first. And it has a great memory. Its antibodies and T-cells, which target bacteria and viruses, are produced in the thymus and bone marrow by a process that resembles the ensemble forecasting approach: you don't know what the intruder will look like, so you try everything. The genes for these proteins are mixed and matched randomly, to generate an enormous sample of different shapes. When one of them matches an intruder, a positive feedback switch is activated and more copies are made in the same shape, at a rate of millions per hour. The swelling in the lymph glands during an infection is caused by the rapid growth of colonies of immune cells. When the infection is cured, the immune cells degrade—except for a few so-called memory cells, which remain and speed the reaction to any subsequent infection. Vaccines work by stimulating the production of such memory cells.

Despite the efficiency of the human immune system, it occasionally loses the battle for prediction. AIDS, caused by the HIV virus, has infected tens of millions worldwide, and over 25 million in Africa alone; tuberculosis (from a bacterium) and malaria (a microscopic parasite) kill millions each year; and new diseases are constantly emerging. Creutzfeldt-Jakob disease (CJD), the human analog to mad cow disease, is caused not by a microbe but by misfolded proteins known as prions. In the 1990s, it was feared that CJD could kill millions of people, though more recent estimates are far lower.[76] The 2003 SARS outbreak was caused by a novel virus that was much less transmissible than influenza, but it still managed to spread from rural China to five countries within a single day. It killed under a thousand, far fewer than a normal flu, but the resulting panic caused an economic crisis in much of Asia and in cities like Toronto, with a total price tag estimated at $30 billion. It showed that we are only a viral mutation or two away from a modern-day plague—but it also showed that the response can be more damaging than the problem itself.

In 2005 and 2006, the world was working itself into a similar panic over a disease that is highly contagious, extremely lethal, and absolutely terrifying—at least for birds. Like the 1918 influenza, which took more lives than the First World War, the H5N1 strain of avian flu is made up mostly of genes from our feathered friends. But unlike the 1918 flu, it has yet to become easily transmissible between humans.[77] Spread around the world by migratory birds such as ducks, which can host the virus without dying from it, avian flu has killed hundreds of millions of chickens and other birds. Of the humans who have contracted it, usually through direct contact with birds, it has killed about half. (This compares with a mortality rate of 2 to 3 percent for the 1918 pandemic.)[78] If it mutates to a different, human-transmissible form, with even half the original virulence, then things would look extremely bad. But as Laurie Garrett, the author of *The Coming Plague*, pointed out in May 2005, "We have no idea what exact genetic changes this would require, how difficult it is for the virus to make those changes and whether or not the virus would significantly sacrifice its virulence level in the process."[79] There is no "normal" size for an outbreak. Avian flu may therefore turn into something far worse than the 1918 epidemic, or it may never happen.[80]

Mathematicians and epidemiologists can model the spread of disease with computer simulations, which in their cruder form divide a population into three classes: those who have the disease, those who are susceptible, and those who are immune (possibly because they have died). The models assume that people encounter each other randomly, like molecules in a gas, and pass on the disease at a rate that depends on its transmissibility.[81] Epidemics, if they become established, are then seen to follow a typical S-shaped curve, rather like that in figure 7.1 (see page 282).

More sophisticated models simulate a detailed population that statistically matches the properties of a given city. Each "individual" in the model will interact with a certain number of people each day. The number of interactions varies just as it does in a real town, so that students attending class come into contact with many people, while those who work at home do not. These simulations have shown that the speed with which health officials act—by isolating patients and telling people to stay at home—is critical.[82] Indeed, the impact of SARS was much lower in Vancouver than in Toronto because health-care workers there immediately managed to contain the disease.

However, the most important factor is the nature of the disease itself, and we cannot predict what exactly will emerge from the global ecosystem.[83] The process by which viruses incorporate new genes is inherently random. The evolution of a microbe as it adapts to a new environment is also highly variable and unpredictable. In one experiment, two samples of a virus that usually infects the bacterium *E. coli* were introduced to *Salmonella* instead. The two virus samples adapted in completely different ways, and after just ten days, they were genetically distinct.[84]

Nor can we predict the lethality or transmissibility of a disease by sequencing a microbe's DNA, since symptoms (like traits) are an emergent feature of the complex interaction between the microbe and its host. This was scarily demonstrated in 2001 by Australian researchers who were trying to develop a mouse contraceptive as a population-control device. Their approach involved the insertion of the gene interleukin-4 into the genome of mousepox (a version of smallpox that afflicts mice but not humans). The gene was expected to boost antibody production, but instead it transformed the virus into a raging killer that took out all the laboratory mice.[85] A major concern is that biotechnologists will accidentally create novel diseases that control the populations of both mice and men.

According to legend, Apollo's arrow enabled Pythagoras to cure plagues, but there seems little that mathematical models can do to offer similar protection. Building a model of an epidemic in progress is rather like working out how to double the cube at the altar of Apollo during the Athenian plague. So how can we prepare ourselves for the emergence of new diseases?

The best option, it seems, is to take a cue from our own immune systems, both innate and acquired. The innate response is to couple systems such as the Global Public Health Intelligence Network,[86] which automatically searches news reports and websites worldwide to pick up signs of an outbreak, with available technologies such as antiviral drugs[87] and antibiotics. The acquired response is vaccination tailored to the particular disease. The World Health Organization is attempting to improve systems for the design, production, and distribution of vaccines.[88] New biotechnologies can play an important role by speeding the process, and high-throughput techniques can monitor the development of dangerous pathogens. Equally important will be low-tech plans to maintain basic health care and food supplies during an outbreak.

Of course, as Timothy Geithner, president of the Federal Reserve Bank of New York, remarked on the slightly different subject of hedge funds, "It is hard to motivate people to buy more insurance against adverse outcomes when the risks seem remote and hard to measure and when present conditions seem favourable."[89] The price tag of the next epidemic is uncertain, but a protective network can be accurately costed.[90] It is natural for those footing the bill to demand proof that the investment is worthwhile. But how can we prove whether avian flu is the revenge of chickens on the human race? As Garrett says, "The bottom line for policymakers: Science does not know the answer."[91] It is hard to get the balance right, and even our own immune system overreacts sometimes. All we can do is watch for coming storms.

WE DON'T KNOW

It might seem in this chapter that we have fallen into the trap of assuming that the future will resemble the past: just because we cannot predict atmospheric, biological, or economic systems now does not mean that we will not be able to do so one day, once we have better computers, observation systems, and models. Statements about the limitations of human ingenuity have a way of being proven wrong. However, the constraints here are the result of the nature of the systems themselves. In cellular automata, computational irreducibility does not (by definition) go away with a better computer, and emergent properties cannot (by definition) be expressed in terms of simple physical laws. Similarly, the dual nature of complex real-world systems means that they can be based on simple, local rules—avian flu is just a few bird genes—but at the same time be uncomputable. Errors in model parameterizations are magnified by the complex feedback loops that characterize such systems. In a hundred years, we won't have an equation for the Game of Life, and (to venture a prediction) we won't have an equation for life.

Even if models do not have predictive accuracy, they are still useful tools for understanding the present, envisaging future scenarios and educating policy makers and the public. Scientific research into global warming—along with pictures of calving icebergs—has highlighted the changes that are currently happening in the climate system, and provoked debate about the possible consequences. Similarly, models of the spread of disease have helped focus people's minds on the possible consequences of a pandemic, and economic models have shown how dependent we are on our relationship with the rest of the biosphere. For, in the long run, what unites our future weather, health and wealth is that they all rely on the state of the planet.

To summarize:

- Prediction is a holistic business. Our future weather, health, and wealth depend on interrelated effects and must be treated in an integrated fashion.

- Long-term prediction is no easier than short-term prediction. The comparison with reality is just farther away.

- We cannot accurately predict systems such as the climate for two reasons: (1) We don't have the equations. In an uncomputable system, they don't exist; and (2) The ones we have are sensitive to errors in parameterization. Small changes to existing models often result in a wide spread of different predictions.

- We cannot accurately state the uncertainty in predictions. For the same two reasons.

- The effects of climate change on health and the economy (and their effects on the climate) are even harder to forecast. When different models are combined, the uncertainties multiply.

- The emergence of new diseases is inherently random and unpredictable. Avian flu may be the next big killer—but a bigger worry is the one that no one has heard about yet.

- Simple predictions are still possible. These usually take the form of general warnings rather than precise statements.

- Models can help us understand system fragilities. A warmer climate may cause tundra to melt and rainforests to burn, thus releasing their massive stores of carbon. However, the models cannot predict the exact probability of such events, or their exact consequences.

- Uncertainty means that discussions become polarized between opposing camps of optimists and pessimists. This, combined with the difficulty in costing unknown future perils, creates a bias towards inaction.

Given that optimists and pessimists are unlikely to convince one another on purely theoretical grounds of our future impact on the planet, and its impact on us, it is hard to imagine how the debate will be resolved until events unfold. To better prepare for an uncertain future, perhaps we need to discontinue the search for more accurate numerical predictions and do the opposite instead. We take up this theme in the next chapter.

8 ▸ BACK TO THE DRAWING BOARD
FIGURING OUT WHERE WE WENT WRONG

Hereafter, when they come to model Heaven
And calculate the stars, how they will wield
The mighty frame, how build, unbuild, contrive
To save appearances, how gird the sphere
With centric and eccentric scribbled o'er,
Cycle and epicycle, orb in orb.
—John Milton, *Paradise Lost*

You don't need a weather model to know which way the wind blows.
—*After Bob Dylan*

UNDER THE VOLCANO

In Autumn 2004, all eyes, along with a large number of highly sensitive scientific instruments, were focused on Mount St. Helens, in the Cascade Mountains. Back in May 1980, Mount St. Helens had been hit by a magnitude 5.1 earthquake. The earthquake had set off a massive volcanic eruption, along with the largest landslide in the earth's recorded history. A plume of gas and ash reached twenty-five kilometres high in less than fifteen minutes. Some 500 million tons of ash were sent into the atmosphere, blotting out the sun for hundreds of miles around; the ash had spread around the world in two weeks. Fifty-seven people died in the blast; some managed to outrace it by driving at speeds of up to 150 kilometres an hour.[1] Now it looked like the mountain might do it again. Trembling to a series of smaller earthquakes, burping the occasional stream of hot ash—was this the prelude to another major eruption or just a passing phase?

Mount St. Helens is part of the Ring of Fire, a network of volcanoes and faults that forms a circle around the Pacific Ocean and marks the boundaries of geological plates whose pressure and shear against each other is balanced by the forces of friction. Any small slip or shudder alters this balance and sends molten rock to the surface. The lava in Mount St. Helens is squeezed up into a six-kilometre-wide chamber. Heat and pressure fracture the rock, turn groundwater and glacial ice in the crater to steam, and push up against the lava dome, a cap of cooled magma that acts like a cork in a bottle of champagne.

Scientists from around the world camped out near the mountain to monitor its movements. Small aircraft circled like flies, with onboard devices sniffing for signs of gases that might indicate the magma was rising. Global-positioning instruments and seismometers picked up minute movements and tremors, and microphones near the crater listened for any sign of activity.

If anything was being more closely monitored and recorded than the mountain, it was the musings of the scientists on the evening news. The question every reporter from around the world wanted answered was, Will the mountain blow? The answers were always guarded and usually uninformative. There could be a major eruption, but it's unlikely. The most probable outcome is a minor eruption. Or it could stop now and lapse back into a coma, not saying a word for a hundred years.

In the end, nothing happened. The mountain calmed down. The foreign reporters returned to Japan or France or wherever they were from. Washington State moved its attention back to the upcoming presidential election, and the Ring of Fire seemed snuffed out—until December 26, 2004, when it roared back to life not as a volcanic eruption in Washington, but as a massive earthquake on the other side of the world.

The magnitude 9.0 underwater quake off the west coast of Sumatra was caused when the floor of the Indian Ocean slipped under the Burma plate, creating a seabed cliff over 10 metres high and 1,000 kilometres long. The sudden movement initiated a tsunami that raced across the ocean at 900 kilometres per hour towards the beaches of Indonesia, Thailand, India, and Sri Lanka. As the waves neared the coast, they slowed and grew to a height of several metres before slamming into the land. Their destructive force swept away people on the beach, and in villages and cities. The death toll of nearly a quarter million made it one of the biggest natural disasters in history.

No reporters anticipated this event. Nor were there tidal gauges or buoys to give an early warning. Yet the earthquake had been felt around the world, and within fifteen minutes of its occurence, scientists had informed several countries (including Thailand and Indonesia) that dangerous waves might be generated; however, because such warnings usually turn out to be false alarms, the information was not acted on. As one professor stated, "We have believed as a community that the Indian Ocean is fairly immune to tsunamis of the kind that took place."[2]

PREDICTING EARTHQUAKES

Earthquakes are caused by the slipping and sliding of huge plates of rock that float on the earth's molten core. When the mechanics of plate tectonics were first understood in the 1960s, it heralded an era of optimism about earthquake prediction. Billions of dollars were funnelled into research programs. Japan, which found itself on the intersection of three such plates, spent ¥160 billion over thirty years. Theories abounded that earthquakes followed some predictable pattern or had reliable precursors.

Unfortunately, the research has shown little success. Earthquakes, like the one that devastated the Kashmir area of India and Pakistan in October 2005, are as unpredictable today as they were thirty years ago. No precursors have been identified, and most of the research programs have been wound up.

The problem is that earthquakes represent a shift in balance between two opposing forces: the dynamical force that grinds one plate against another, and the frictional force that resists movement. Tension is released not in a smooth, continuous manner, but in sudden fits and starts, felt as tremors or quakes. The situation is in some respects similar to collapses in financial markets, which are the result of a sudden shift in balance between buyers and sellers. Like financial crashes, earthquakes occur at unpredictable intervals, and their magnitude tends to follow a power-law distribution: there are many smaller ones, and fewer large ones.

Even if the timing of earthquakes eludes prediction, it is possible to forecast where they are most likely to occur from the location of the tectonic plates. Engineers can also calculate how manmade structures will react and design building codes for areas at risk. Whether their advice is heeded is another question—it hasn't been in much of northern California, which rides along the San Andreas fault, part of the Ring of Fire.

We would all like to know what is on the horizon for the earth and sky over the next one hundred years. Are we living under a volcano? Will global warming or other human causes trigger a major disaster, or will the threat fizzle out like Mount St. Helens? Yet our best models of weather, health, and wealth regularly fail to predict even short-term phenomena. So perhaps an equally important question is, What are we doing wrong? Is there something fundamentally askew in our mental approach, rather than just the technical execution? Are we looking for answers in the wrong place? In this chapter, we put our three sibling oracles on the analyst's couch to find out why they keep going astray—and whether their behaviour can be blamed on the parents.

CAUSE AND EFFECT

In the nineteenth century, the psychologist Franz Brentano observed that people divide the world into two categories and handle each in different ways.[3] The first category consists of things that act spontaneously and with intentionality—in short, are alive. The second includes things that obey physical laws. The distinction between the two classes is not necessarily for complicated philosophical or religious reasons, or because we assign a mysterious "vital force" to one and not the other, but is due to the simple empirical fact that they behave differently. Living beings have evolved in a way that gives them special properties. Kick a stone, and a sense of physics will explain what happens; kiss a person, and it's more complicated.

Human beings excel at two types of prediction, and these correspond to the categories above. One is based on empathy—working out what someone else is feeling, putting ourselves in the shoes of another—and the second on cause and effect. The former works best for living beings, the latter for objects. But different people prefer different approaches. Psychologists Simon Baron-Cohen and Alan Leslie have proposed that autistic children find it hard to empathize. As babies, they make less eye contact with their mothers, and later in life, they have problems communicating or maintaining social relationships. In one study, the eye movements of adult autistics were tracked while they watched emotionally charged scenes from the 1966 film *Who's Afraid of Virginia Woolf?*[4] The patients tended to focus on peripheral objects rather than on the eyes of the actors. During a scene in which Richard Burton and Elizabeth Taylor kiss passionately, one viewer kept his attention on a light switch. "The world of objects is much more central to them than the world of people," said Fred Volkmar, one of the study's authors.[5] Risk factors for autism include being male, and having mathematicians or physicists in the family.

Autism, or its weaker form, Asperger syndrome, need not be a barrier to success, or to genius. Hans Asperger, the Austrian doctor for whom the latter condition was named, believed that "a dash of autism" was essential for high achievement in many fields, in particular the "highly specialised academic professions, with a preference for abstract content."[6] It has been speculated that some of the greatest scientists of all time, including Albert Einstein, Isaac Newton, and even Socrates, showed classic signs of having Asperger syndrome.[7] Newton, for example, was famously anti-social and incommunicative. If no one showed up for his lectures, he just gave them to the empty room.[8]

EXTREME SCIENCE

In June 2000, following a lecture at the Sorbonne by the economist Bernard Geurrien on the disconnect between conventional economics and reality, a group of fifteen French students unleashed a kind of primal scream against their educational establishments. In a petition signed by hundreds, they exclaimed: "We wish to escape from imaginary worlds! . . . We oppose the uncontrolled use of mathematics! . . . We are for a pluralism of approaches in economics! . . . Call to teachers: wake up before it is too late! . . . We no longer want to have this autistic science imposed on us."[9] Thus was born *autisme-économie,* or the post-autistic economics (PAE) movement. An interview in the French newspaper *Le Monde* garnered the students instant attention; the movement has since spawned many research papers, books, and a journal.

The choice of name was unfortunate—the students' intention was not to criticize people with autistic disorders. However, the comparison between mainstream theory and an extreme mental condition makes sense, and not just in economics. A similar "extreme" pattern is evident in the archetypes and structures of Pythagorean thought. It is perhaps best summarized by the Pythagorean's list of opposites in table 1.1 (see page 29), which like a set of aesthetic principles has shaped all three areas of prediction.

ONE VS PLURALITY—As the physicist B. K. Ridley wrote, "Mathematical physicists are motivated by the vision of Oneness. They are offended by the plurality of particles on the one hand and by the plurality of forces on the other."[10] Many genome scientists, in their quest for genetic causes of complex traits, appear to be driven by what the biologist Richard Lewontin describes as the "ideology of simple unitary causes."[11] The economist Frank Knight wrote, "This is the way our minds work; we must divide to conquer. Where a complex situation can be dealt with as a whole—if that ever happens—there is no occasion for 'thought.'"[12]

RIGHT VS LEFT—Pythagorean science emphasized rational, logical thought (since Hippasus, "irrational" has been the worst term of abuse in science).[13] This is a speciality of the left hemisphere of the brain, which controls the right side of the body. Both hemispheres of the brain work in unison, but loosely speaking, the left tends to use a predictive approach to problem solving, based on mental models and causality, while the right uses an integrative approach, which takes into account context and the big picture. In one experiment, split-brain patients were shown a sequence of red and green dots, which is random but biased so reds appear 75 percent of the time and greens only 25 percent. It was found that the right hemisphere tends to guess the next dot will be red (the simple climatological approach) while the left hemisphere will search for a complicated but nonexistent pattern, and do less well.[14]

MALE VS FEMALE—As Evelyn Fox Keller noted, modern science was developed "not by humankind but by men."[15] The first secretary of the Royal Society, Henry Oldenburg, described its aim as being "to raise a Masculine philosophy."[16] Similarly, Jay Griffiths wrote that mechanistic science has stripped away "chance, caprice and unpredictability—all things which, for good and for bad, have been associated with the female."[17] The economist Julie A. Nelson said that "analytical methods associated with detachment, mathematical reasoning, formality, and abstraction have cultural associations that are positive and masculine, in contrast with methods associated with connectedness, verbal reasoning, informality, and concrete detail, which are culturally considered feminine."[18] The oracles are still male.

AT REST VS IN MOTION—the Pythagoreans resisted any idea of movement in mathematics apart from the fixed rotation of stars and planets, which spin like tops in their crystalline shells. Copernicus wrote: "We conceive immobility to be nobler and more divine than mutability and instability."[19] The aim of predictive science is to find permanent, immutable laws. It is married to progress, but in the sense that it wants to fill in that single fixed image to greater and greater detail. As Galileo pointed out, however, the difference between mutability and immutability is "exactly the difference between a living animal and a dead one."

STRAIGHT VS CROOKED—Mechanistic science has long been based on linear cause-and-effect relationships, for example, by drawing straight lines through clouds of data to deduce correlations. The emphasis on linearity began to change in the 1960s—weather models are certainly not linear—but even now, the crooked is ignored wherever possible (as Mandelbrot found when he tried to apply his fractal geometry to economics). The reason: straight lines are predictable, while crooked lines weave and change.

ODD VS EVEN—The Pythagoreans associated the even numbers with the dyad two, which signified mutability, excess, conflict, and indeterminacy. Mechanistic science is similarly intolerant of duality and mutability.

LIMITED VS UNLIMITED—The goal of mechanistic science is always to fix and constrain nature, to make it predictable. Even in chaos theory, much emphasis has been placed on finding attractors that limit the system to a small band of motion. Since Adam Smith and Malthus, the "dismal science" of economics has been structured around the ideas of scarcity and limits.

LIGHT VS DARKNESS—Mechanistic science tries, like Kepler did, "to draw the obscure facts of nature into the bright light of knowledge."[20] It has illuminated much, but it still often ends up fumbling in the dark for the switch.

SQUARE VS OBLONG—Squares and circles have a deep geometric symmetry, which allows them to be described by a single parameter. Oblongs and ovals spoil this symmetry, as do clouds. Kepler wanted nothing more than to "square" his oval orbits, which he called a "cart-ful of dung."[21] Newton rehabilitated them by showing that it wasn't the shape of the orbit that mattered but the form of the underlying law—and nothing is more symmetrical or invariant than the law of gravity.

Many truly innovative scientists, including Kepler, Newton, and Pythagoras, have mixed rationality with a kind of mysticism that defies categorization. However, it is fair to say that the basic thought processes of mechanistic science lean heavily towards the first column of opposites. There is nothing bad about the first column—it even includes the word "good" (in Good versus Evil). As a practising mathematician, I have spent a lot of time there myself. I enjoy clean lines and modernist architecture as much as the next person. But one can have too much of a Good thing. There has emerged a worldview that, while in many respects tremendously successful—we could say it got us where we are today—is also one-sided and extreme. The Pythagorean quest for the harmony of the spheres did not end with Kepler, but has been channelled into mechanistic models that reduce the natural world to a collection of simple objects that can be understood and controlled.

In fields such as complexity, systems biology, or theoretical physics, scientists have in recent decades begun to move towards a rounder, more holistic, "post-Pythagorean" perspective. By necessity, they have had to grapple with duality, with mixing the left and the right.[22] Engineers have always been ready to adopt top-down approaches. Predictive scientists, however, still cling to a purely mechanistic model of the world—not because they are stuck in the past or have been programmed by an ancient Apollonic cult, but because to do otherwise is to admit that the world is not predictable. They view the world as a collection of inert things—objects rather than subjects.

SUBJECT OR OBJECT

In 1967, the physicist Erwin Schrödinger observed that science is based on two beliefs: that nature is both objectifiable and knowable. The first means that we can isolate ourselves from nature and study it as an independent object from a position of lofty detachment. As the philosopher Mary Midgley noted, the emphasis on objectivity has had a profound effect on social scientists, who "have often pursued a very powerful and confused notion of 'objectivity' as requiring, not just the avoidance of personal bias, but a refusal to talk or think about subjective factors at all. The word 'subjective' then becomes a simple term of abuse directed at any mention of thoughts or feelings, and the word 'objective' a potent compliment for any approach which ignores them."[23] The result is that even human beings are treated as lifeless objects. Unlike the blind Gloucester in Shakespeare's *King Lear*, a scientist should not see the world feelingly.

Weather and climate modellers too have always been afraid of what the meteorologist Carl-Gustaf Rossby called the "horrible subjectivity." The word "forecast" was invented by Robert FitzRoy, precisely to avoid the subjective connotations of "prediction" and any unwelcome comparisons with Zadkiel's Almanac. But changing the name did not make the issue of subjectivity go away.[24]

The Greek Circle Model, with its swirling epicycles, maintained its hold on the human imagination for 2,000 years not just because of its prominent use of circles, which according to Ptolemy were alone in being "strangers to disparities and disorders," but also because it worked as a predictive device (it could accurately forecast eclipses and the movements of the planets). A thirteen-year-old Tycho Brahe became fascinated by astronomy when he witnessed a partial eclipse of the sun that had been accurately foretold. It seemed "something divine that men could know the motions of the stars so accurately that they were able a long time beforehand to predict their places and relative positions."[25] Christopher Columbus wowed the natives of Jamaica into submission by accurately predicting a lunar eclipse on February 29, 1502, then promising to return their moon if they obeyed him.[26] Einstein's theory of relativity was accepted not because a committee agreed that it was a very sensible model, but because its predictions, most of which were highly counterintuitive, could be experimentally verified. Modern GCMs have no such objective claim to validity, because they cannot predict the weather over any relevant time scale. Many of their parameters are invented and adjusted to approximate past climate patterns.[27] Even if this is done using mathematical procedures, the process is no less subjective because the goals and assumptions are those of the model builders. Their projections into the future —especially when combined with the output of economic models—are therefore a kind of fiction. The fact that climate change is an important and contentious issue makes it all the more important that we acknowledge this. The problem with the models is not that they are subjective or objective—there is nothing wrong with a good story, or an informed and honestly argued opinion. It is that they are couched in the language of mathematics and probabilities: subjectivity masquerading as objectivity. Like the Wizard of Oz, they are a bit of a sham.

Schrödinger's second tenet—that nature is knowable—implies there is a kind of inbuilt correspondence between our minds and nature. Just as Pythagoras and Kepler believed the motion of the planets revealed a cosmic harmony that could be understood by mathematics, modern scientists see the climate system, and even life itself, as a challenging but tractable mathematical problem. It is indeed remarkable how fundamental forces of nature, such as gravity, seem to conform to mathematical equations; but the same is not necessarily true of complex emergent properties, such as life. As Evelyn Fox Keller noted, "Belief in the knowability of nature is implicitly a belief in a one-to-one correspondence between theory and reality."[28] This affects the kind of problems that scientists pose. "Questions asked about objects with which one feels kinship are likely to differ from questions asked about objects one sees as unalterably alien. Similarly, explanations that satisfy us about a natural world that is seen as 'blind, simple and dumb,' ontologically inferior, may seem less self-evidently satisfying for a natural world seen as complex and, itself, resourceful."[29]

So what difference would it make to long-term predictions for the planet if the planet itself were treated as a living entity?

IT'S ALIVE

The idea that the earth behaves like a self-regulating organism is the essence of James Lovelock's Gaia theory. In the early 1960s, before Lovelock used his electron-capture gas chromatograph to detect the buildup of ozone-eating CFCs in the atmosphere, he was invited by NASA to invent a different kind of instrument—one capable of being sent on a spacecraft to answer that famous question, posed by David Bowie, "Is there life on Mars?" The first problem was what to test for, since any life forms on Mars may be radically different from those on earth. The most general characteristic of life, it seemed, was that it takes in energy and matter and discards waste products. Such processes should leave a chemical signature on the Martian atmosphere.

To test his idea, Lovelock and Dian Hitchcock began to analyze the chemical makeup of Mars and compare it with that of the earth. The results showed a strong contrast. The atmosphere of Mars, like that of Venus, was about 95 percent carbon dioxide, with some oxygen and no methane. The earth was 77 percent nitrogen, 21 percent oxygen, and a relatively large amount of methane. Mars was chemically dead; all the reactions that were going to take place had already done so. The earth, however, was far from chemical equilibrium. Methane and oxygen, which react with each other very easily, are both present in the atmosphere. Lovelock concluded that for this to be the case, the gases had to be in constant dynamic circulation, and the pump driving this circulation was life.[30]

About 3 billion years ago, bacteria and photosynthetic algae started to remove carbon dioxide from the young earth's atmosphere, producing oxygen as a waste product. Over enormous time periods, this process changed the chemical content of the atmosphere—to the point where organisms began to suffer from oxygen poisoning. The situation was relieved only with the advent of other life forms powered by aerobic consumption. It was life processes, the cumulative actions of countless organisms, that made the atmosphere we now enjoy. The blanket of greenhouse gases that controls our temperature, the ozone layer that acts as sunscreen, the entire physical condition of the planet—all have been actively shaped by life itself. The net effect of these processes was that the earth itself appeared as a living entity—a kind of super-organism. As Lovelock wrote, "An awesome thought came to me. . . . Could it be that life on Earth not only made the atmosphere, but also regulated it—keeping it at a constant composition, and at a level favourable for organisms?"[31] On a stroll with his novelist neighbour, William Golding, Lovelock described his idea and asked advice for a name. Golding suggested naming it after Gaia, the Greek earth goddess (who was displaced by Apollo from the seat of prophecy along with Sybil and Python).

The Gaia hypothesis hinged on the observation that the planet is self-regulating, or homeostatic. In an extension of Darwinism, life forms are regulated by the environment, and in turn the environment is actively regulated for life. As we saw in Chapter 5, homeostasis is not the same as static behaviour—all life forms are in a constant state of flux—but refers instead to the ability to retain a degree of internal order in a changing environment. The heat of the sun has increased by 25 percent since life began on earth, yet the temperature has remained more or less constant (on this scale, even ice ages count as relatively minor fluctuations). The human body regulates its temperature by shivering if too cold or perspiring if too hot, and other mechanisms. Lovelock, together with the American microbiologist Lynn Margulis, uncovered a number of feedback loops that could act as regulatory influences on a planetary scale. These are similar to those discussed in Chapter 7, but they operate over longer geological time scales.

An example is the long-term regulation of carbon dioxide. When volcanoes such as Mount St. Helens erupt, they throw massive quantities of carbon dioxide into the atmosphere. If the carbon were allowed to build up over millennia, the greenhouse effect would make the earth too warm to support life. One process by which carbon dioxide is removed from the atmosphere is rock weathering, where rainwater and carbon dioxide combine with rocks to form carbonates. Lovelock, Margulis, and others discovered that this process is hugely accelerated by the presence of soil bacteria. The carbonates are washed into the ocean, where algae use them to make microscopic shells. When the algae die, their shells sink to the ocean floor, forming limestone sediments. Since the soil bacteria are more active in high temperatures, the removal of carbon dioxide is accelerated when the planet is hot. This has the effect of cooling the planet. Therefore, the whole massive cycle forms a stabilizing negative feedback loop. The idea of self-regulation was illustrated with the conceptual model Daisyworld (see Appendix III).

One feature of such feedback loops is that they combine living and non-living components; another is that they make the climate extremely hard to model. No perturbation has a straightforward, linear effect; the response is always crooked. The regulatory systems that Lovelock and Margulis saw as a sign of life are also a sign of unpredictability when viewed by a mathematical modeller. The multiple non-linear feedback loops create sensitivity to errors in parameterization (see figure A.5 in Appendix III, page 359). A corollary to the Gaia hypothesis could state that the robustness of the earth system is inversely related to our ability to predict its future.

Is the Earth Alive?

Does it make sense to view the world as a living organism? One can argue that the planet is only a physical system with biological components; it has no mind, soul, free will, or DNA, so it can't be alive. But on the other hand, plants don't have minds, animals aren't supposed to have souls, not everyone believes we have free will, and the earth has plenty of the rather inert substance known as DNA, including our own. There are many definitions of what constitutes life, and some would include the earth system.

One useful definition, proposed by the Chilean neuroscientists Humberto Maturana and Francisco Varela in 1987, states that a being is alive if it is autopoietic, or self-making—it produces the components that define it as a unit. What is important in this definition is not so much the material structure of life but the process, organization, and relationships of the components. For something to be alive by this definition, there is no requirement that it grow or reproduce or pass on DNA.

Other definitions assume that organisms reproduce and don't recycle their own waste, so the earth system would not be alive. Perhaps the earth can be viewed as the opposite end of the spectrum from viruses: the latter cannot reproduce except through a host body; the earth is a host body that cannot reproduce (except perhaps through colonization). You can argue about their status as living beings, but to make predictions you have to take them both into account.

TELLING STORIES

The idea that the earth is a kind of super-organism is a very old one. As Leonardo da Vinci wrote, "By the ancients man has been called the world in miniature; and certainly this name is well bestowed, because, inasmuch as man is composed of earth, water, air and fire, his body resembles that of the earth; and as man has in him bones the supports and frameworks of his flesh, the world has its rocks the supports of the earth; as man has in him a pool of blood in which the lungs rise and fall in breathing, so the body of the earth has its ocean tide which likewise rises and falls every six hours, as if the world breathed; as in that pool of blood veins have their origin, which ramify all over the human body, so likewise the ocean sea fills the body of the earth with infinite springs of water."[32] This ancient idea gained special resonance in December 1968, with Apollo 8: that space mission allowed the earth to be viewed for the first time as a complete entity from outer space, so obviously alive in comparison with its barren neighbours, the moon and the other planets.

While Gaia theory is now well established in ecological science, and has helped spawn interdisciplinary fields such as earth system science (a kind of planetary-scale systems biology), it has sometimes been seen as bad or even "dangerous" science.[33] Part of the reason may be its overt use of metaphor: we can't prove objectively that the earth is alive; it is just a useful way of seeing things. As Lovelock wrote, metaphor is seen by scientists "as a pejorative, something inexact and therefore unscientific."[34] Aristotle said that "metaphor is a poetic device, but it does not advance our knowledge of nature." Scientists like to chastise the media for caring only about catchy stories (at least, those the scientists don't agree with). But models, in the end, are always stories, metaphors for the real world. They explain a sequence of events; they are products of the mind. The only way to avoid metaphor is to limit oneself to the pure forms of mathematical abstraction.

Science is therefore charged through with metaphor and mythology. The clockwork universe is a metaphor, and so are the butterfly effect, the selfish gene, and the efficient market. Compare, for example, the latter's deification of the marketplace, with its ability to look far into the future to calculate value,[35] with the psychologist James Hillman's definition of the "archetypal premise in Apollo" as "detachment, dispassion, exclusive masculinity, clarity, formal beauty, farsighted aim and elitism."[36] Perhaps the efficient market hypothesis should be renamed the Apollo hypothesis. Most of the early opposition to Gaia theory came from proponents of the selfish gene theory, which managed to ascribe Apollonic properties to a section of chemical code. The butterfly effect allows all short-term error to be ascribed to chaos, leaving the GCMs untouched as the paragons of reason and far-sighted accuracy. Like the clockwork universe, these catchy metaphors all enforce a vision of the world as a rational, computable machine; they emphasize one side of the story. They encourage a kind of resignation in the face of determinism, as if we live in a world without choices: we are the victims of the initial condition/our genes/the market.[37] The danger occurs when practitioners adopt these stories unconsciously and are unaware of their subtext.

The real issue, then, is not so much that Gaia is a metaphor, but that it is the wrong kind of metaphor. The science writer Margaret Wertheim compares the Pythagorean philosophy to a mythological Earth Mother/Sky Father polarity: "In seeking to free the immaterial psyche from the material body, the Pythagoreans were seeking to escape from the realm of the Earth Mother (in their mathematical mythology, this being represented by the number 2), and to ascend into the realm of the Sky Father (represented by the number 1)."[38] From this point of view, Gaia theory will only hinder union with the divine harmony. Climate modellers probably don't think about their jobs in quite those terms, but stories, myths, and metaphors can be as powerful and enduring as mathematical theorems. Like any mental model, they affect what we allow ourselves to see and the problems we attempt to solve.[39]

If the earth is more like a self-regulating organism than a piece of rock floating through space, then that is important information that has implications for our future.[40] The questions we ask, and the kind of predictions we make, must change completely. The biosphere has evolved in a way that makes it hard to predict.[41] That mix of gases that has been in the atmosphere since pre-industrial times isn't there by accident; it's part of a biological system, and that affects what will happen if we perturb it. And as environmentalists have often pointed out, our species may resemble nothing so much as a bad infection or a cancer.[42] The main distinction between diseased and healthy states is a degree of excess, and our population has expanded until it has taken over much of the planet, with its impact on the environment multiplied many times by technology. We have commandeered natural systems, turning forests and rivers to our own use the way a tumour takes over blood supplies. Our economy's toxic waste is poisoning the planet and upsetting the very regulatory networks that sustain its homeostasis. Our growth is out of control.

When any organism is threatened, it fights back. Robustness does not imply passive tolerance. Our bodies combat infection by raising the temperature, and global warming might also turn into something akin to a fever. Epidemics might similarly be seen as an antibody-like defence mechanism designed to keep us in check. As Lovelock warned, "We are at war with the Earth itself."[43] Viewed in this way, climatological, biological, and economic prediction must be seen as three aspects of the same thing. This explains why it's so difficult to make long-term predictions of our future weather, health, and wealth: they all depend on the reaction of a robust, living planet under serious assault. This isn't a calculation; it's a medical crisis.

THE CASSANDRA COMPLEX

Of course, the idea that mankind is a disease on the planet is just another story. We are always reassessing our role: we used to be the apple of our creator's eye, then we were another ape, and now we're a disease. It is also unnecessarily fatalistic: we have choices and can control our future. Our nature is not fixed, and disaster is not in our genes. However, if we accept that our actions are affecting key regulatory networks in the earth system, then any prognostication about the future becomes—metaphorically—like a prognosis for a patient. Medics can cover the earth with instruments and stick thermometers in every orifice and monitor each subtle symptom as it appears, but it's hard to work out from any of these what the future holds. The normal procedure is to check medical history to identify the disease and discover its usual course; but in this case, it's a new condition. The earth is like the first victim to stumble into a hospital emergency room. We've never seen *Homo sapiens* on this scale before. How to respond? An overly technical doctor might want to order up new scans using the latest equipment; a penny-pinching hospital administrator might say it's all in the patient's mind.

The current debate about global warming—between scientists who believe that global warming is an important issue and "skeptics" who don't want to make large economic sacrifices for a problem that might not exist—resembles the argument between the technician and the administrator squabbling in the E.R. corridor. The technician has the scientific expertise and the support of most of the staff, but he is still not winning. The administrator has all the good lines, his arguments are well organized, and he's louder. It's an entertaining debate, and the other staff members gather round to watch. In the room behind the door, though, the patient is feeling worse—there are complications.

As is often the case in debates between people holding two extreme positions, they are both right, and they are both wrong. The technician is right to say that he believes the planet's life systems are under threat; he is wrong to believe he can predict the outcome using technology. The administrator is right to insist that the technician's analyses have been inaccurate in the past and are not a reliable guide to the future; he is wrong to insist on absolute proof when it is clearly impossible. They both view the patient as an object, either medical or economic.

Most people, including most politicians, are neither climate scientists nor global-warming skeptics; yet we in the industrialized world still tend to see the world in objective terms, as something to be manipulated and controlled, slave to the laws of cause and effect. This is the shadow side of our great inheritance from two millennia of science. Those predictive models of the world not only looked into the future but, in many respects, helped define it. By turning the world into an object that we can control, however, we also deny it life. And by closing off our emotional involvement with nature, we become unable to take the necessary decisions to protect it. We might be willing to buy the environmental argument intellectu-ally, and it does seem to be getting warmer out. But as any good consumer knows, decisions are driven less by logic than by feel-ing. There is an intense drama going on (*Who's Afraid of the Planet Earth?*), but we are looking for the light switch. We tend to get more emotional about the killing of baby seals in the Canadian North than the melting of the Canadian North; more worked up about second-hand smoke from cigarettes than that from the exhaust pipes of cars; more worried about risk factors for disease than the health of the planet. A Gallup poll in April 2006, for example, con-cluded that Americans are "still not highly concerned about global warming," with no increase since the question was first surveyed in 1989, even though more agree that it is happening.[44]

The failure of our forecasting models, and the ancient dream to mathematically predict and control the future, grows out of this confusion between objects and living things. The authors of the Club of Rome's World3 program, for example, say that formulating their ensemble predictions for the planet's future is like throwing a ball up into the air. "To predict exactly how high the ball will rise or precisely where and when it will hit the ground, you would need precise information. . . . Therefore we put into World3 the kinds of information one uses to understand the generic behaviour modes of thrown balls, not the kinds of information one would need to describe the exact trajectory of one particular throw of one specific ball."[45] But the world is not an inert ball, and there is no generic response. Perhaps that is why, like Cassandra, such models so often fail to convince or to move.

Lack of predictability is a deep property of life. Any organism that is too predictable in its behaviour will die. And in an unpredictable environment, the ability to act creatively, while maintaining a kind of dynamic internal order, is a prerequisite. The balance of positive and negative feedback loops, when combined with the computational irreducibility of life processes, makes the behaviour of complex life forms impossible to accurately model. The problem is not that such organisms are erratic, but that they combine creativity with control. House plants are quite stable (they tend to stay in their pots and don't suddenly walk off to join the forest), but it would still be impossible to predict the exact effect of moving a single plant from a shaded spot to a warm greenhouse, based only on a detailed understanding of its biochemistry. If we can't do it for a plant, we can't do it for a planet. Life, it seems, evolves towards rich, complex structures, which defy simplistic analysis.

Acknowledging the liveliness of the earth's response changes the questions we ask. While we once wanted to know when our needs would be met, we now ask, Is this working out? Are we doing our part? Are we close to crossing a line? And how do we feel about that? To answer these questions, we need better metaphors, better stories.[46] Like an overly formulaic Hollywood film, our current

9 ► CONSULTING THE CRYSTAL BALL
OUR WORLD IN 2100

But what have been thy answers, what but dark
Ambiguous, and with double sense deluding,
Which they who asked have seldom understood . . .
No more shalt thou by oracling abuse
The Gentiles; henceforth oracles are ceased,
And thou no more with pomp and sacrifice
Shalt be enquired at Delphos or elsewhere,
At least in vain, for they shall find thee mute.
   —John Milton, *Paradise Regained*

Let now the astrologers, the star-gazers, the monthly
prognosticators, stand up and save thee from these things that
shall come upon thee.
   —Isaiah, 47:13

FORECAST 2100

So now that we have all those theoretical points and disclaimers out
of the way, we can ask how things will really look in the year 2100.
While researching this book, I came across a variety of ideas, scenarios, predictions, and concerns. Most are based on the output of

GCMs, coupled in some cases with models of physical, biological or economic systems. Others are speculations based on what appear to be credible scenarios. The most plausible are listed below.

- The average global temperature will rise by about five degrees (C or F).
- Droughts in places such as Spain, Australia, New Zealand, the Middle East, and parts of the United States will make it difficult to grow traditional crops.
- Wheat yields will improve in Canada and Russia.
- Sea levels will rise by a metre or more.
- Summer monsoons in Asia will be more variable, with increased risks of floods or droughts.
- Three million cubic kilometres of ice in the Greenland ice sheet will begin a long and unstoppable melting process.[1]
- The West Antarctic ice sheet will also begin to melt.[2]
- Glaciers worldwide will continue to recede.
- The Arctic will have ice-free summers, impacting on ice-living animals, birds, and northern indigenous peoples.[3]
- Much of the tundra in northern countries will disappear, releasing its stores of carbon.
- A combination of fires and pest outbreaks will severely damage boreal forests in China and other countries.[4]
- Huge dust storms in the Gobi and Sahara deserts will cause respiratory problems worldwide.[5]
- Local warming and rainfall reduction will cause parts of the Amazonian rainforest to collapse and die, releasing their stores of carbon.
- Wetlands such as South America's Pantanal will dry out, impacting species such as migratory birds.
- Storms and hurricanes will dramatically intensify.[6]

- Areas including France, Germany, and the northwest United States will experience increased heat waves, like the one that hit Paris in the summer of 2003.[7]
- Coastal erosion will displace hundreds of millions of people, destroy prime farmland, flood entire island nations, and result in huge costs for cities such as Alexandria, Amsterdam, Manila, Calcutta, and London.
- The thermohaline ocean circulation will slow or stop, causing the U.K. winter to go Canadian.[8]
- Warmer oceans will result in quasi-permanent El Niño conditions.
- Exhausted fisheries will not recover.[9]
- Coral reefs will turn white.[10]
- Losses in species diversity will result in widespread ecosystem collapse.[11]
- Global warming will accelerate disease spread in a range of species, from coral to Hawaiian songbirds.[12]
- Dengue fever, malaria, and other mosquito-borne tropical illnesses will head north.
- The increased incursion of humans into natural habitats will bring new and deadly diseases.[13]
- Biotechnologists will accidentally or deliberately create novel pathogens that will be released into the population.[14]
- Our increased population density, coupled with rapid transportation networks, will result in fast-spreading pandemics.
- The gap between rich and poor will accelerate, leading to increased social and economic instabilities.
- The NASDAQ stock index will reach one million.[15]
- Poor people will cluster in vulnerable areas, and the number of lives lost to natural disasters will continue to climb.
- Wars will erupt over water, as well as oil.[16]

- Local shortages of food and water will lead to mass migrations.
- Climate disruption, unsustainable land use, ecosystem collapse, population growth, pollution, and other factors will combine to reinforce one another and accelerate the degradation of the planet.
- An asteroid at least fifty kilometres wide will collide with the earth sometime during the century, killing millions of people.[17]
- The release of tiny, self-replicating machines invented by nanotechnologists will reduce the surface of the planet to a "grey goo."[18]
- There will be a nuclear war, followed by a nuclear winter.[19]
- Civilization will collapse globally.

Other things to look out for are the following:

- The average global temperature will be little changed.
- The growth in global communication, coupled with increased interest in the environment, will help slow or reverse environmental damage.
- People will switch to fuel-cell cars or hybrids or bicycles or foot-power or public transit or stay at home in huge numbers.
- Countries will become far more efficient in energy use, embracing non-carbon energy sources such as solar, biomass, wind, etc.
- Countries will switch to nuclear reactors, perhaps based on fusion technology.[20]
- Most cancers will be curable or treatable in rich countries.
- Stem-cell and other genetic treatments will extend the lifespans of wealthy people.
- Nanomachines will help control global warming by removing carbon from the atmosphere (before they convert the earth's surface to a grey goo).

- We will rapidly evolve taboos against pollution and over-population.
- Partly as a result of a booming global economy, birth rates will fall more quickly than anticipated. The earth's population will not be much greater than today's.[21]
- Carbon emissions will stabilize.
- We will experience a revolution on the scale of the agricultural and industrial revolutions.[22]
- Civilization will prosper globally.
- The NASDAQ stock index will cease to exist.
- The earth system will recover quickly from the damage we are doing.
- The earth system will cure itself, in a way that is bad for us.[23]

Finally, we might:

- begin to see the planet as a living system, and as a result stop damaging it
- denounce our oracles as false, deluding, and distracting—or simply stop listening to them

Except perhaps for the last, these predictions are consistent with GCM forecasts and IPCC projections under different economic scenarios, but represent only a sample of the known unknowns. There are, of course, also the unknown unknowns.

Given that we can rule none of these out, how can we determine the probability of any of them happening? Is the chance that global warming will be greater than 2°C just 10 percent, or is it more like 90 percent? What exactly are the odds?

Robert FitzRoy defined a forecast as something completely objective, "the result of a scientific combination and calculation."

But as I argued in this book, we cannot obtain accurate equations for atmospheric, biological, or social systems, and those we have are typically sensitive to errors in parameterization. By varying a handful of parameters within apparently reasonable bounds, we can get a single climate model to give radically different answers. These problems do not go away with more research or a faster computer; the number of unknown parameters explodes, and the crystal ball grows murkier still. There is no State of Civilization Risk Assessment Test (SOCRATES) to tell us the answer. We can't mathematically calculate the odds, even if it looks serious, scientific, and somehow reassuring to do so. Like Socrates himself, we only know that we know nothing. Any prediction necessarily involves a large dollop of subjectivity. Numerical models are not enough.

One way forward is to ask experts what they think. This is the approach of the IPCC, who asked themselves. Based on "collective judgment of the authors, using the observational evidence, modeling results, and theory that they have examined," the IPCC concludes, for example, that there is a 90 to 99 percent probability of "increased heat stress in livestock and wildlife" because of higher maximum temperatures, but only a 67 to 90 percent chance of "increased damage to coastal ecosystems such as coral reefs and mangroves" because of more intense tropical cyclones.[24] Such fuzzy estimates are a useful way to present complex information and attempt to rank threats. However, the results are panel-dependent, and not everyone shares the IPCC's robust belief in the models. A group of economists would probably conclude that in a hundred years, we and our livestock will all be living in air-conditioned bubbles, and a panel of skeptics would insist that we will acclimatize to slightly warmer conditions. A panel consisting of Astronomer Royal Sir Martin Rees, meanwhile, would give us only a 50 percent chance of surviving at all, which puts things into perspective.[25]

To get a balanced assessment, we have established a panel of enlightened, rational, far-sighted experts—the true paragons of reason. Yes, we mean you, the readers of *The Future of Everything*. Please join the online forum at www.apollosarrow.ca to vote on issues from the likely extent of global warming and the chance of a global pandemic to the probability of a major shift towards a sustainable economy. We will tabulate the responses in real time and use them to formulate a prediction. The result will be carefully archived. It may not predict the future, but will at least provide amusement for anyone from future generations who happens across it.

GREAT PREDICTIONS FROM HISTORY

No book on prediction would be complete without a salute to famous predictions from yesteryear.

"If one might trust the Pythagoreans, who believe in the recurrence of precisely the same series of events, you will be sitting there, and I shall be holding this staff and telling you my story, and everything will be the same." —Eudemus, a student of Aristotle, commenting (around 300 B.C.) on the Pythagorean notion of eternal recurrence. That would explain those faint feelings of déjà vu.

"Whatever befalls the Earth, befalls the sons and daughters of the Earth." —Chief Seattle, 1854. Not a prediction so much as a warning.

"When the Paris Exhibition closes, electric light will close with it and no more be heard of." —Erasmus Wilson, Oxford University, 1878. Electric lights are now visible from space.

"The inhabitant of London could order by telephone, sipping his morning tea in bed, the various products of the whole earth and reasonably expect their early delivery upon his doorstep; he could at the same moment and by the same means adventure his wealth in the natural resources and new enterprises of any quarter of the world, and share, without exertion or even trouble, in their prospective fruits and advantages." —John Maynard Keynes predicting Internet commerce, around 1900.

"The horse is here to stay, but the automobile is only a novelty." —The president of the Michigan Savings Bank advises against investing in the Ford Motor Company, 1903.

"Sooner or later a crash is coming, and it may be terrific . . . . Factories will shut down . . . men will be thrown out of work. . . . The vicious circle will get in full swing and the result will be a serious business depression." —Roger Babson, 1929. His remarks came before the worst crash in U.S. stock-market history—and may have helped trigger it.

"I think there is a world market for maybe five computers." —Thomas Watson, 1943. A good thing for this chairman of IBM that he was wrong.

Computer chips should double in power at a rate of "roughly a factor of two per year." —In 1975, Gordon E. Moore modified his 1965 prediction (otherwise known as Moore's Law) to a doubling every two years. It has proved quite accurate. If computer-based predictions had kept pace since the 1950s, they would have improved by a factor of a million, and we would be able to see into the next century.

"By the year 2000, people will work no more than four days a week and less than eight hours a day. With legal holidays and long vacations, this could result in an annual working period of 147 days worked and 218 days off." —From the *New York Times,* October 19, 1967. If only we could get away from those computers.

"Genetic control or influence over the 'basic constitution' of an individual." —First predicted in 1967 to be available by the year 2000, it was assessed by a panel of scientists in 2002 as "a prediction that might occur, but hasn't happened yet."[26]

"There is now considerable evidence that the first stages of the next ice age may really begin soon, within the next few years—and that the transitional stage of extreme and inhospitable climate may already have begun." —Larry Ephron in *The End: The Imminent Ice Age and How We Can Stop It* (1988). Global warming put a stop to that.

We are "moving toward the greatest worldwide depression in history, in which millions of people will suffer catastrophic financial reversals. . . . It will occur in 1990 and plague the world through at least 1996." —Ravi Batra, talking up what became a huge economic boom, in *The Great Depression of 1990* (1987).

"You can live long enough to live forever." And see the future yourself. An optimistic (I think) 2004 forecast from the anti-aging guru Ray Kurtzweil. So long as you remember to look each way before crossing the road, from now until the end of time.

THE FUTURE FORETOLD (FOR NON-MATHEMATICIANS ONLY)

In the meantime, allow me to probe the tangled entrails, the whispering trees, the babbling brooks of subjectivity for omens and portents and wholesome advice. . . .

I feel we're living in a bubble, but I can't prove it, and I can't call the top. I think, at a global level, we won't run out of energy, water, or resources anytime soon, but the planet still has limits, and just because they bend (they are not fixed or immutable) doesn't mean they won't break. I think we will affect the climate, and not be best pleased with the warm and erratic outcome. But the real problems will be down here on the ground, not up there in the sky. I think we won't be able to produce detailed and convincing predictions of the changes before they happen. I think we should have fewer children, pollute less, tread more lightly—and not wait for our scientists to compute us a solution. It's a time for action, not calculation. I feel a storm is coming, but I don't know if it is atmospheric, medical, economic, or all three. I feel we're in only the second act of this particular story—tension is building, forces are aligning, clouds are gathering—and it's not clear what twists lie ahead or how matters will resolve. I think nature has a few tricks left up her sleeve.

I think this is pretty unoriginal (it's not rocket science), but I can't prove it objectively, rationally, or mathematically, because it's not that kind of problem. Life is not a predictable machine. Life is a surprise.

But . . . when Kepler asked his tutor, Mästlin, for advice on astrological predictions, he told him just to predict disaster, since that was bound to come true sooner or later. So a disaster: sometime in the next hundred years, just when overpopulation and environmental stress seem to be the biggest problems, and many in poorer countries are weakened by drought and famine, there will be a worldwide pandemic. Our global, interlinked, just-in-time economic system will fall apart as countries impose quarantines and people stay at home. A couple of years later, when the disease has run its course, we will try to start up the economic machine again—but rust will have set in. Carbon emissions will decline, and the climate will eventually stabilize. After a period of wars, invasions, and insurgencies, so will we. Life will return once again to normal, with the difference that we are wiser, more humble, and more respectful of nature.

And then we'll do it again.

Or, scenario B—and here I embrace the non-objective, ensemble-forecasting approach—we get a warning shot that we can't ignore. This kick-starts a third, already nascent revolution, one that's of the same magnitude as the agricultural and industrial revolutions but does not involve a new way of extracting energy from the ground. We'll know it after we see it. Carbon emissions will decline, and the climate will eventually stabilize.

DEFENCE OF SOCRATES

As we've seen in this book, mathematical models have consistently failed to provide accurate predictions of atmospheric, biological, or economic systems—they do not know the future. Such a statement usually has negative connotations: "I don't know" is associated with failure, bad marks on exams, and in Socrates' case, the sipping of poisonous hemlock drinks. But it does not mean that mathematical models are of no use in addressing the world's problems or understanding the present. Our impact on the planet can be visualized only with scientific technology that extends our senses to a global level. The important thing is that we do not allow an unbalanced and often fake insistence on objectivity to distance us from the world or cut off our connection to it. The future depends on the choices we make, and on the reactions of complex systems that are beyond our control. Decision-making in such situations relies as much on felt cultural and political values as it does on logical analysis. Objectivity and subjectivity must be in balance, and inform each other, just like the positive and negative feedback loops that characterize living systems. We will choose to protect nature only if we value it—and not just as an object, but because it is alive. The only way we will respect it is if we understand that we cannot control it.

In non-linear, complex systems, change often happens abruptly, like water turning to ice. Extreme change is normal. This makes prediction difficult, but it also holds out tremendous hope, because it means that a sudden change in course can be expected. Such change often comes from the bottom up, rather than from the top down; it comes as a felt reaction, rather than something told to us by experts. Unlike deterministic mechanical systems, we have a choice; we can determine our own destiny. We are not slaves to the initial condition, our genes, or the efficient market. We are unpredictable, and that's no bad thing.

The science of complexity will not build a better GCM, and neither will Gaia theory or earth system science. Their stories are more of humility than of human ingenuity. But if we as a species are standing at a precipice, it is better that we see the world feelingly than be completely blinded by our mental models; that we know what we do not know.[27] Creativity often emerges from a state of uncertainty. Grasping for illusory knowledge by over-modelling our environment is therefore part of the problem.

Even if we cannot predict storms, we can predict our ability to weather them. Engineers can calculate the vulnerability of structures to disasters such as flooding, hurricanes, and tsunamis, and help design suitable building codes. Economists can point out weaknesses in a country's financial system, and health workers can determine how much medicine will be needed to fight an epidemic. But as our populations extend into floodplains and coastal areas, and global warming raises sea levels and increases the force of storms, we may find that the walls we built to withstand those once-in-a-lifetime storms are no longer high enough to keep the water out.

Mathematical models will always be indispensable. Like language, they are a way to understand the world, and organize and communicate our thoughts. They help us perform hypothetical experiments, explore possible scenarios, and expose fragilities. Most of all, they help us comprehend what is happening now. The mathematician Ralph Abraham wrote: "While we may not be able to predict the future with certainty, or at all, we may at least exercise our cognitive processes, with mathematical models and computer graphic simulations that improve our understanding of the present, enhancing our chances of survival in the future."[28] Apollo's arrow cannot fly into the future or protect us from plague, but it may serve as a compass, point out dangers, and help us navigate an unpredictable world.

Appendices

These appendices present three conceptual models that illustrate some of the ideas in the book. The first is based on bread-making, the second on the flow of air, the third on the growth of daisies. Each tells a simple story, like a fable, and has its own moral.

Appendix I: The Shift Map

The shift map was introduced in Chapter 3 as a simple example of a chaotic system. The dynamics, which are similar to the kneading of bread dough, were illustrated in figure 3.2 (see page 100). The first few iterations, for two nearby initial conditions, are shown in the left panel of figure A.1. The solid and dashed lines could represent the positions of two yeast cells in the dough, on a scale from 0 to 1. They start close together, but after just a few iterations, they are completely separated.

Now, suppose that we wish to predict the future location of the daughter cell (dashed line in the left panel), based on the position of the mother cell (solid line). Over the first few steps, the error will increase exponentially. The solid line in the right panel shows the average error growth over a large number of experiments from different starting points, for the same small initial error of 0.005. The average separation indicates what we could expect the prediction error to be after a certain number of iterations. As seen, the exponential growth eventually saturates at the average distance between two randomly chosen points. This is ⅓, or about 0.33.[1]

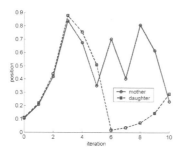 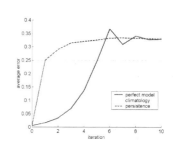

FIGURE A.1. The left panel tracks the location of two yeast cells in dough (on a scale of 0 to 1), separated by 0.005. The distance grows by a factor of two for the first few iterations, then becomes more random. The right panel shows the average or expected error, over a large number of experiments from different starting points, created by a small perturbation to the initial condition. It grows exponentially at first, then saturates at a level equal to the average distance between randomly selected points. The best results for long-term prediction are given by the climatology model, which uses the centre of the dough as its forecast and has an expected error of only 0.25. The dashed line shows error for a persistence forecast, which assumes that the cell stays where it started.

After only a few steps, any prediction performed by applying the shift map is no better than random. A better guess is to use as our "model" the long-term average of all possible states, which is 0.5. This so-called climatological forecast has an average error, for all points including the first, of 0.25, as shown by the dotted line. The dashed line shows error for a persistence forecast, which assumes that the cell stays where it started. For the shift map, we could say that the perfect model (i.e., using the shift map) is best over the short term, while the climatological model is best over the long term. The persistence forecast has no advantages.

The shift map can be represented mathematically as a sequence of transformations. Given an initial condition x_0, the next iteration is $x_1 = \text{mod}(2x_0)$, where the mod function means that if $2x_0$ is greater than 1, only the fractional part is used. This process is repeated, so the nth term in the sequence is given by $x_n = \text{mod}(2x_{n-1})$. The shift map is so named because its effect is to shift the binary expansion of the starting point one place to the right. Any point can be represented in a binary expansion as a string of zeros and ones. For example, ½ is 0.1 in binary notation; ¼ is 0.01; and ¾ (i.e., ½ + ¼) is 0.11.[2] Since computers work in zeros and ones, all numbers are converted to binary notation internally. If the initial point in binary notation is $x_0 = .11010110\ldots$, where the higher digits are omitted, then after one step the point will have moved to $x_1 = .1010110\ldots$, $x_2 = .010110\ldots$, and so on.

Another way to view the shift map is therefore as a simple one-dimensional cellular automaton.[3] The binary representation of any initial condition can be represented as a line of cells, with white cells corresponding to a 0 and black cells corresponding to a 1. The top line of figure A.2 reads .11010110 . . . , and the lines below show that each iteration just shifts the cells to one side.

FIGURE A.2. Black cells represent 1, white cells represent 0 in a binary expansion. The top line shows the initial condition .11010110 . . . , which is shifted one cell to the left at each iteration of the map.

Our impression of this system depends very much on point of view. If we concentrate on the magnitude of the difference between two points, then we see sensitivity to initial condition and chaotic behaviour. However, if we view the map as operating on a binary expansion, it is just a line of squares marching along, with one falling off the end at each iteration. The system itself is therefore completely predictable. Anyone can compute the effect of a thousand iterations just by shifting the binary representation a thousand places. The only unpredictability—the lack of order and form that the word "chaos" evokes—is a result of error in the measurement of initial condition.

MORAL: Chaotic systems are not the same as complex systems. In the former, prediction requires an accurate initial condition; in the latter, prediction is impossible.

APPENDIX II: THE LORENZ SYSTEM

As discussed in Chapter 4, this system was used by Ed Lorenz from MIT in 1963 to illustrate the effect of chaos on predictability.[4] The system is an approximate model of the convective motion of a fluid heated from below and cooled from above, a problem first studied by the Victorian physicist Lord Rayleigh. The differential equations, which express the rate of change of the three variables, are:

$$dx/dt = -\sigma x + \sigma y$$
$$dy/dt = -xz + Rx - y$$
$$dz/dt = xy - Bz$$

The parameters are usually set to $\sigma = 10$, $B = 8/3$, and $R = 28$. The equations were based on a seven-equation model by Barry Saltzman, which itself was truncated from an infinite series of terms derived from a two-dimensional approximation. The system therefore represents an extreme simplification of the actual dynamics.

As shown by the dashed line in figure A.3, the system is sensitive to initial condition: root mean square errors caused by a small change in the initial condition increase in a quasi-exponential fashion with time (as in figure A.1). More significant, though, is the error caused by a 4 percent change in the single parameter R. The relative importance of initial condition and parameter errors depends on the details (and can be measured using the drift technique discussed in Chapter 4), but the point is that the system is sensitive to both. Error cannot necessarily be pinned on the initial values of variables, rather than the parameters. The error in R corresponds to a similar error in the assumed temperature differential of the two plates. In models of the real weather, model error arises not just because of a single parameter, but because the equations themselves cannot reproduce the detailed behaviour of the atmospheric system.

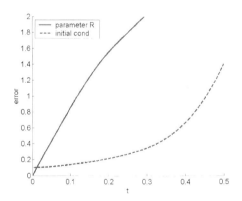

FIGURE A.3. Plot showing root mean square errors caused by a small change of magnitude 0.1 in the initial condition (dashed line) and a 4 percent change in the parameter R from 28 to 29.12 (solid line). Short-term predictions are sensitive to both initial conditions and parameters. As in the right panel of figure A.1, errors eventually saturate.

Suppose now that the goal is to predict the long-term "climate" of the Lorenz system, in terms of the variance of the fluctuations in x, y, and z. Over the long term, the initial condition is no longer important, because the system settles onto its own attractor; however, the climate depends critically on the parameters and the model equations. A 4 percent increase in R makes the butterfly attractor grow in size, so the variance in each of the three variables increases by about 5 percent—a significant change, since it increases the probability of extreme events.[5] The attractor of Saltzman's original seven-equation model has a completely different appearance—the butterfly grows a body and looks more like a moth.[6]

The Lorenz system is therefore an example of a chaotic system that is sensitive to changes in parameters and equations when either short- or long-term prediction is the goal. As discussed in Chapter 7, models of the real climate show similar properties.

MORAL: In the long run, and often the short run too, errors in equations matter more than initial errors in the variables—even in chaotic systems.[7]

APPENDIX III: DAISYWORLD

Daisyworld is a mathematical model that illustrates self-regulating, homeostatic behaviour and represents a simple example of an ecosystem. It was initially developed by James Lovelock and Andrew Watson as a defence of Gaia theory.[8] This theory was based on the observation that the earth system seemed to regulate itself like a biological organism. Perhaps taking the idea a little too literally, some mistakenly interpreted it as a claim that the earth was behaving with a sense of purpose, that it was a teleological being actively controlling the climate and so on.[9] The purpose of Daisyworld was to show that homeostatic behaviour could emerge naturally, without recourse to teleology. It also exhibits a number of other properties that are typical of models of living systems.

On Daisyworld, an imaginary planet, there are only two species of life: light daisies and dark daisies. Light daisies tend to reflect light, which has a cooling effect, while dark ones absorb radiation, and therefore warm the planet. Growth of the daisies depends on the present population, the natural death rate, the available space, and the temperature. The equations are based on the dynamics of real daisy growth. Details about the equations, as well as animated simulations, can be found at a number of websites.[10]

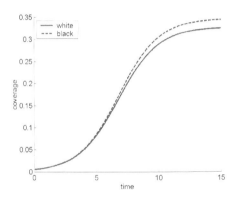

FIGURE A.4. Plot showing the growth of white and black daisies, from an initial seeding. The proportion of white and black daisies at equilibrium acts to regulate the planetary temperature.

Figure A.4 shows a simulation of the growth of the two species, from the initial seeding to their equilibrium level. (The model was also used to obtain the S-shaped curve for economic growth in figure 6.1 on page 237.) The growth can be described in three parts. At first, the seeded daisies grow exponentially, multiplying rapidly and establishing themselves on the planet. In the second phase, growth is complicated by lack of space and competition with other daisies—negative feedback. Finally, in the third phase, the daisies obtain a balanced equilibrium. Life processes still go on, but the number of births is balanced by the number of deaths, so there is no net growth. More complicated versions produce chaotic fluctuations around a mean.

The aim of the model was to illustrate the idea of planetary homeostasis—in particular, how the earth could regulate its temperature as the sun's luminosity slowly increased. The Daisyworld planet revolves around a sun, absorbing energy at a rate that depends on the sun's luminosity and the albedo of the planet. It also radiates heat out to the universe, just as the earth does. Black daisies absorb heat and do best in cool conditions. When the planet is cool, they grow more quickly. But in doing so, they lower the albedo of the planet as a whole, thus warming the planet. Similarly, if the planet gets too hot, it will be cooled by the growth of white daisies, which (like polar ice) reflect heat. The two species therefore act antagonistically. Since they are also in competition for space, growth in one tends to repress growth in the other. The final temperature represents a compromise between the two types of daisies—neither too hot, nor too cold. When the model is run with the sun's luminosity gradually increasing, the light and dark daisies adjust themselves naturally to keep the temperature constant at the optimal level for daisy growth.

Daisyworld is therefore an example of a self-regulating system, where internal conflict between opposing forces somehow conspires to maintain the conditions suitable for life. From the outside, it seems that little is happening; the daisies of different species carry on their lives in balance with each other. Yet stability can be deceptive. Powerful forces are at work to maintain that tranquility in the face of external changes. As Heraclitus wrote: "An invisible harmony is stronger than a visible one."[11] One consequence is that the Daisyworld "climate" is insensitive to changes in initial conditions but sensitive to changes in parameters. This is illustrated in figure A.5, which shows the reaction to different kinds of perturbations. The vertical axis represents the temperature of the white daisies, which has a steady state indicated by the square symbols (as mentioned above, more complex versions produce fluctuations around a mean; homeostasis implies regulation but not static behaviour). If the initial daisy population is increased by 10 percent, competition for space means that the excess daisies die off and equilibrium is rapidly restored (circle symbols). The rate of return depends on the rates of growth and decay, and the steady state represents a balance between these opposing forces. A smaller 2 percent change in a single parameter, which controls the death rate of the white daisies, is enough to create a significant shift in the balance between the two species and affect the temperature (diamond symbols). Note that this does not represent some defect in the equations, but instead is a natural property of the system itself. As an archer takes aim, his two arms similarly pull in opposite directions to tension the bow, and balance each other without conscious direction on his part. The bow is at rest, and from the outside it again looks as if little is happening. But any mathematical model of his movements would be unstable to small random errors in the simulation of the force applied by either hand.

The yeast galactose model discussed in Chapter 5 has some similar dynamics, with the two species of regulatory proteins playing the role of daisies. The rapid growth/decay rates of these species, combined with their opposing regulatory effects, results in a homeostatic system that is highly resistant to both external perturbations and internal stochastic noise (but sensitive to changes in parameters). Since most of the parameters could only be estimated to within about a factor of ten, they had to be carefully balanced against one another to make the model work.

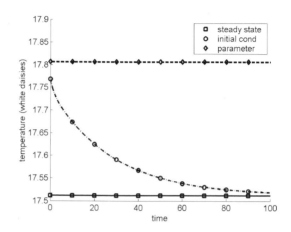

FIGURE A.5. Plot showing sensitivity to parameterization but insensitivity to initial condition, in the Daisyworld system. The normal system has a steady state attractor (square symbols), and a perturbation of 10 percent in the initial condition is rapidly damped out (circle symbols). However, a 2 percent error in one of the parameters results in a new and significantly different steady state (diamond symbols). The reason is that the steady state represents an equilibrium between opposing forces, and it's easily disturbed by a small error modelling one of them. Note that in most biological systems, it is impossible to measure parameters to anything like 2 percent accuracy.

Daisyworld is only a kind of thought experiment, but it shows how self-regulation can occur without any recourse to teleology. Species in an ecosystem can be concerned with nothing more than their own survival, yet they help not only themselves but the whole system. Daisyworld opened people's eyes to similar regulatory loops in our own planet. An example is the salinity of the oceans, which is about 3.4 percent.[12] Tears from your eyes are about as salty, and most living organisms maintain a salinity that is roughly equal to that of the oceans. Perhaps this is why Empedocles proposed the metaphor that the sea is the sweat of the earth, which so annoyed Aristotle.

One theory about salinity was that natural selection tended to assist those organisms that were in balance with their surroundings. So why has the ocean managed to maintain a constant level of salinity? If it were to go much above 4 percent, then basic cell functions, such as the electromagnetic potential maintained across cell membranes, would fail. There would be mass extinctions of life in the oceans. And yet there is no evidence of such extinctions in the past 500 million years, despite the fact that salt is constantly being deposited in the oceans through the weathering of rocks. Furthermore, there have been cataclysmic events—meteorite impacts, periods of glaciation, and so on—that one might expect to abruptly alter salinity. Indeed, attempts to model the salinity regulation using chemistry or physics have failed. So what is regulating the oceans?

From Daisyworld, we might predict that the answer is the organisms that live there. In fact, bacteria play a particularly important role in the running of the oceans (as they do in most life processes). Although they constitute only 10 to 40 percent of the ocean biomass, they make up 70 to 90 percent of the biologically active surface area. And they all pump salt. Looking at the problem from the point of view of Gaia theory breaks down the barriers between what we have traditionally seen as living and non-living systems.

Moral: Self-regulating systems, which consist of a dynamical balance between opposing forces, are hard to model and predict.

Notes

► **INTRODUCTION**
THE SCIENCE AND SOCIOLOGY OF FORECASTING

1. McCarthy 2004.
2. Gunson 1999.
3. Gutkin 1999.
4. Wieczorek et al. 2001.
5. Ibid.
6. Gutkin 1999.
7. In 1999, a survey by the Pan American Health Organization showed that 30.6 percent of Caracas residents had been victimized by crime that year; the statistics were probably higher in the *ranchos*.
8. In 1798, flash floods damaged over 200 homes and government buildings and knocked out all the bridges. Geologists have also uncovered a record of floods going back to prehistory. Röhl 1950.
9. See Trenberth 2005 for a discussion of the effect of global warming on hurricanes; see also Webster et al. 2005.
10. Brookes 2004, p. 133.
11. Galton 1878.
12. They were all men. As a colleague pointed out, the few women in the field for some reason seemed more receptive to the idea of model error.
13. Ray and Anderson 2000, p. 273.
14. "Despite the appearance of analytical rigour, IIASA's widely acclaimed global energy projections are highly unstable and based on informal guesswork." Keepin and Wynne 1984.
15. Mandelbrot and Hudson 2004, p. 167.
16. Kuhn 1962.

17. The peer-review process, which decides what research will be published, has its benefits but is also political. *The Economist* notes, "When the peers in question are drawn from a restricted professional domain . . . peer review is not a way to assure the highest standards of work by exposing research to skepticism. It is just the opposite: a kind of intellectual restrictive practice, which allows flawed or downright shoddy work to acquire a standing it does not deserve." As everyone knows, peer pressure makes you conform, but it doesn't make you right, especially when the peers are a small clique. Anonymous 2003. See also Keepin and Wynne 1984.

18. As Bart Kosko has pointed out, scientists like to control what is said about them in the press. See Kosko 1993, p. 42; see also Saul 1992, p. 79.

19. See Tetlock 2005 for a study of political predictions; also Sherden 1998.

20. Sunnafrank and Ramirez 2004.

21. Science also aims to explain the present, but theories are not usually considered satisfactory unless they can make verifiable predictions. This is in some respects a recent development. See Oreskes 2000.

22. See, for example, Fischetti 2001.

23. Keller 1985.

1 ► SLINGS AND ARROWS
THE BEGINNINGS OF PREDICTION

1. Ovid, *Metamorphoses.* Translated by Sir Samuel Garth, John Dryden, et al. Online at http://classics.mit.edu.

2. Powell 1998, p. 172.

3. Abraham 1994, p. 92.

4. Wood 2003.

5. The gas might have been the narcotic ethylene, released through faults in the rock. Roach 2001.

6. Cornford 1969, p. 35.

7. Iamblichus 1918, p. 6.

8. Ibid.

9. His followers often suggested that Pythagoras was Apollo in human form. See Strohmeir and Westbrook 1999 for a general history of Pythagoras. The exact dates for the life of Pythagoras are not known. See Guthrie 1962, p. 173.

10. Iamblichus 1918, p. 49.

11. Ibid., p. 50.
12. Strohmeir and Westbrook 1999, p. 49.
13. Iamblichus 1918, p. 50.
14. Ibid., p. 11.
15. Ibid., p. 21.
16. Miller 2006.
17. From Bogen 1975, quoted in Edwards 1999, p. 37.
18. Ibid.; Guthrie 1962, p. 252.
19. Iamblichus 1918, p. 79.
20. Midgley 1985, p. 98. This is a shortened version of her table.
21. Guthrie 1962, p. 458.
22. Iamblichus, 1918, p. 59.
23. Ibid., p. 32.
24. To see that the square root of 2 cannot be written as a rational number A/B, suppose first that it can, so $(A/B)^2 = 2$. We can assume that A and B do not have a common factor, otherwise this can be divided out. Multiplying each side of the equation by B^2 gives $A^2 = 2B^2$. The number A^2 is even, since it is a multiple of 2. Therefore A must also be even (because if it was odd, the square would be odd). Since A is even, we can write $A = 2C$ for some C. So substituting into $A^2 = 2B^2$ gives $(2C)^2 = 2B^2$. Therefore $4C^2 = 2B^2$, or dividing each side by 2, $2C^2 = B^2$. Since the left-hand side is even, it follows that B^2 is even, and therefore that B is even (if it was odd, the square would be odd). So A and B are both even and have a common factor of 2. But this contradicts the assumption that A and B do not have a common factor. The only way out is to conclude that the initial assumption is wrong; in other words, the square root of 2 cannot be written in the form A/B.
25. The dart would have to be infinitely thin to pick out exactly one number. In mathematical terms, the measure of the irrational numbers in the line segment from 0 to 1 is 1, while the measure of the rationals is 0.
26. See Guthrie 1962, pp. 176–77.
27. Koestler 1968, p. 25.
28. Plato, *Apology*. Translated by Benjamin Jowett. Online at http://classics.mit.edu. Seaton 2004.
29. Field 1956. Guthrie also writes that "Plato's debt to the Pythagoreans is obvious." Guthrie 1978, p. 216.
30. Thucydides 1910. The disease may not correspond to a modern one.

31. It was believed in ancient times that plague was caused by Apollo's arrows. For this reason, St. Sebastian, who is usually pictured with his body run through by arrows, was one of the saints most often invoked. Ferguson 1954.

32. Eudoxus was called a Pythagorean. He was an associate or friend of Plato, but probably not a student. Guthrie 1978, p. 447.

33. Wood 2003, p. 136.

34. Quoted in Peterson 1993, p. 37.

35. Quoted in Van der Waerden 1974.

36. Ibid., p. 48.

37. Reagan 1989.

38. Quoted in Gill et al. 1981, p. 1.

39. Aristotle, *On the Heavens*, 306a. Online at http://classics.mit.edu.

2 ► LET THERE BE LIGHT
TYCHO BRAHE AND THE MODEL MAKERS

1. Watson 1968, p. vii.

2. Quoted in Zöllner and Nathan 2003, p. 104.

3. Koestler 1968, p. 192.

4. Tycho Brahe, *Astronomiae instauratae mechanica* (1598). The Smithsonian Institution has the text and figures online in its digital library at http://www.sil.si.edu.

5. Quoted in Christianson 2000, p. 52.

6. Image from Tycho Brahe, *Astronomiae instauratae mechanica* (1598).

7. Iamblichus 1918, p. 42. From Pope's 1720 translation of *The Iliad*, Book 16, "The Death of Patroclus":

> A Dardan youth there was, well known to fame,
> From Panthus sprung, Euphorbus was his name;
> Famed for the manage of the foaming horse,
> Skill'd in the dart, and matchless in the course

8. See illustration in Christianson 2000, p. 119. From another of his poems:

> You, Ptolemy, Alfonso, Copernicus,
> I gave a hand: you slipped, but I held fast.

The motion of the stars you could not grasp
As I have done. In truth, my work was great:
New pillars raised for heaven's sparkling dome.
Quoted in Christianson 2000, p. 216.

9. Connor 2004, p. 79.
10. Quoted in Koestler 1968, p. 242.
11. Deacon 1968.
12. Iamblichus 1918, p. 17.
13. From *Mysterium Cosmographicum,* by J. Kepler, 1621.
14. Ridley 2001, p. 105. The full title is *A Forerunner to Cosmographical Treatises, containing the Cosmic Mystery of the admirable proportions between the Heavenly Orbits and the true and proper reasons for their Numbers, Magnitudes and Periodic Motions,* by Johannes Kepler, Mathematicus of the Illustrious Estates of Styria, Tübingen, *anno* 1596.
15. Iamblichus 1918, p. 45.
16. Koestler 1968, p. 265.
17. Letter to D. Fabricius, 18 December 1604. Quoted in ibid., p. 330.
18. Vincenzo Galilei, 1581. *Dialogo della musica antica e moderne* (A dialogue on ancient and modern music).
19. Koestler 1968, p. 388.
20. Quoted in Gribbin 2002, p. 56.
21. Connor 2004, p. 360.
22. Quoted in Machamer 1998, pp. 64–65.
23. Galilei 1967, pp. 58–59.
24. The course of the plague through London was dramatized in Defoe's 1722 *A Journal of the Plague Year.*
25. Brumfiel 2004.
26. Wertheim 1995, p. 124.
27. Keller 1985, p. 53.
28. Griffiths 2004, p. 167. Gender here refers to abstract properties, like yin and yang; individuals, of course, have a mix of such characteristics.
29. Iamblichus 1918, p. 36.
30. The solution is obtained by integration and gives $x = -gt^2/2 + v_0 t + x_0$, $v = -gt + v_0$, where x_0 is the initial position and v_0 is the initial velocity of the stone. The plot of x versus t is a parabola, the plot of v a straight line.
31. Gribbin 2002, p. 111.
32. The difference between velocity and speed is that velocity has a

direction (it is a vector), while speed is a magnitude. Here, velocity is a negative number because it is directed down, but speed is never negative.

3 ► DIVIDE AND CONQUER
THE GOSPEL OF DETERMINISTIC SCIENCE

1. Descartes 1960, p. 50.
2. Pierre Simon Laplace, *Essai philosophique sur les probabilités,* 1814. Reproduced in the *Oeuvres complètes de Laplace,* vol. 11 (1886).
3. Quoted in Koestler 1968.
4. Bragg 1999, p. 163.
5. Darwin 1905, p. 1.
6. Ibid., p. 39.
7. Quoted in Hirsch 1984.
8. Ibid.
9. Wertheim 1995, p. 120.
10. Quoted in Sherden 1998, p. 203.
11. Griffiths 2004.
12. Latitude could be determined at sea from the stars, but since they rotate from east to west, knowing the exact time was an essential part of the deduction. Jardine 1999, p. 160.
13. For example, the acceleration of a falling stone is given by $-g$. The effect of wind resistance could be modelled by adding a term such as $-0.1v$, so the acceleration is reduced in magnitude as speed increases. The plot of either of these equations as a function of v gives a straight line: the first is just a constant, so gives a horizontal line, while the second has a slope. Wind-tunnel studies, however, have shown that a better model has a term in v^2 rather than v. The plot of such a function gives a parabola.
14. Yorke and Li 1975.
15. Quoted in Wertheim 1995, p. 250.
16. Capra 1983.
17. Dawkins 1976.
18. Brooks 2002, p. 173.
19. Bohm 1974, pp. 127–28. Quoted in Midgley 1985, p. 50.
20. Quoted in Sherden 1998, p. 225.

21. In 1939, though, Einstein had written to President Roosevelt to encourage him to build the bomb before the Germans. See Saul 1992, p. 305.
22. See, for example, http://mathworld.wolfram.com/Life.html.
23. Wolfram 2002, pp. 715–17.
24. Ibid., p. 381.
25. Ibid., p. 741.
26. Coupland 2002, p. 122.
27. Quoted in Peterson 1993, p. 96.
28. Greenblatt 2004, p. 305.
29. Quoted in Capra 1996, p. 19.
30. For a history of the role of women in science, see Wertheim 1995.

4 ► RED SKY AT NIGHT
PREDICTING THE WEATHER

1. Frisinger 1977, p. 55.
2. Moran et al. 1997, p. 110.
3. Ibid., p. 135.
4. Cox 2002, p. 80.
5. Ibid., p. 81.
6. Gillham 2001, p. 149.
7. Ibid., p. 76.
8. Bjerknes 1904.
9. Cox 2002, p. 161.
10. Walker 1918.
11. Walker's statistical methods were initially met by "extreme skepticism" by the meteorological community. Katz 2002.
12. Abraham 1994, p. 210.
13. Cox 2002, p. 203.
14. Ibid., p. 202.
15. Manabe et al. 1965.
16. Deacon 1968, p. 242.
17. Edwards 2000.
18. National Academy of Sciences, Committee on Atmospheric Sciences Panel on Weather and Climate Modification 1966, pp. 65–67.
19. The shift map, discussed in Chapter 3, has both fixed points, which

are mapped to themselves, and periodic cycles. The point 0, for example, is mapped to 0 in perpetuity, so it is a fixed point. However, any small perturbation will double in size repeatedly under the map, so 0 is not a stable point and not an attractor. Similarly, the point ⅓ is mapped to ⅔, which is mapped back to ⅓, so this forms a periodic loop, which is again unstable and not an attractor.

20. The RMS has a number of mathematical and statistical properties that make it easier to work with than the average. See glossary for definitions of technical terms.

21. Wolfram 2002, p. 381. See Appendix II.

22. The metric was 500 mb, discussed below.

23. Toth et al. 1996. As Wolfram wrote, "It is usually assumed that—like in the Lorenz equations—the phenomenon of chaos must make forecasts that are based on even slightly different initial measurements eventually diverge exponentially." Wolfram 2002, p. 1177.

24. As Lorenz put it, "now we had an excuse." Quoted in Gleick 1987, p. 18. See also Calder 2003, pp. 137–38.

25. Toth et al. 1996.

26. For information on the Earth Simulator Center, see http://www.es.jamstec.go.jp/esc/eng/.

27. See http://www.oso.noaa.gov/goes/.

28. Model output cannot be taken literally (see Roebber et al. 2004). Human forecasters can increase skill by 10 to 15 percent (see Kerr 2004a).

29. Estimate is from the reinsurer Swiss Re. Anonymous 2005f.

30. Rabier et al. 1996.

31. See, for example, Kerr 2004a. Statements such as "Today's three-day forecasts are as good as one-day forecasts were in 1981" (Anonymous 1999b) refer to the 500 mb metric.

32. There is always some turbulence at higher altitudes owing to clouds or large-scale flows like the jet stream, but it is much smaller than at lower altitudes.

33. Assuming that the forecast is free or of negligible cost. Roulston et al. 2003.

34. Results will vary depending upon the area. For an unbiased discussion, see Sherden 1998.

35. Bosart 2003; Ebert et al. 2003; Roebber et al. 2004. The National Weather Service in the U.S. publishes statistics online: see, for example, "Annual HPC -vs- NWP Guidance Threat Scores 1993–2004" at http://www.hpc.ncep.noaa.gov/html/hpcverif.shtml.

36. Kerr 2004a; Roebber et al. 2004.

37. Naïve forecasts can do quite well too. See Sherden 1998.

38. See http://www.nhc.noaa.gov/verification/ and http://www.nhc.noaa.gov/HAW2/english/forecast/errors.shtml. Predicting the number of storms in a given season isn't much easier. On May 16, 2005, NOAA's prediction for the 2005 Atlantic hurricane season (which began on June 1) was for twelve to fifteen tropical storms. On August 21, halfway through the season, the range was updated to eighteen to twenty-one. The final tally was twenty-eight. They attributed the increase to naturally occurring, multi-decadal variability. (See http://www.noaanews.noaa.gov/stories2005/s2438.htm.)

39. Emanuel 2005; Webster et al. 2005.

40. Chagnon 2000.

41. Kerr 2004.

42. As Toth et al. (1996) put it, "The ensemble strategy will work only if the models are good enough that model-related errors do not dominate the final error fields." A similar assumption was made at ECMWF: "From its inception, the EPS [Ensemble Prediction System] has been based on the premise that medium-range forecast errors are predominately associated with uncertainties in initial conditions." Buizza et al. 2000.

43. As Mandelbrot noted in the context of financial prediction, "If you are going to use probability to model a financial market, then you had better use the right kind of probability." Mandelbrot and Hudson 2004, p. 105. The same applies for weather models. Parameters can be varied using a Bayesian approach, which accounts for prior (subjective) beliefs. However, if the model is structurally deficient, there need be no correct value or range of values. The parameter is a property of the model, not of the system. Oreskes et al. 1994; Smith 2000; Orrell 2005a.

44. If a distribution is desired, random errors can be added based on past forecast errors. See, for example, Roulston et al. 2003, figure 7, panels c) and d); and Jewson et al. 2002. One tool for measuring the accuracy of probabilistic forecasts is the Brier score. See Rosenthal 2005, p. 240 for a simple description.

45. Sherden 1998.

46. From a collection of historical American documents at http://www.ku.edu/carrie/docs/amdocs_index.html.

47. Sherden 1998, p. 14.

48. This was pointed out by Lenny Smith. See Matthews 2004.

49. Holton 1992, p. 117.

50. Wolfram 2002, p. 997.

51. Moran et al. 1997, p. 125.

52. Algae may play a role in regulating cloud cover. See Charlson et al. 1987.

53. Jaenicke 2005.

54. Roebber et al. 2004.

55. Thelen and Smith 1994, p. xix. Quoted in Moore 2002, p. 152.

56. Toth et al. 1996.

57. Anonymous reviewer, 2000. Commenting on Orrell et al. 2001.

58. See Junger 1997.

59. Lenny's LSE website is http://stats.lse.ac.uk/smith/.

60. Smith 1996, 1997, 2000; Gilmour 1998. See also Danforth and Yorke 2005.

61. Gilmour 1998 looked at shadowing in a weather-related context.

62. The drift technique was proposed in Orrell 2001, and is developed in Orrell 2005a, Orrell 2005b. The method can be used to separate out the contributions to total forecast error of model error and observation error. The observation error in this case corresponded only to the small truncation error that occurs when translating from the high-resolution model to the low-resolution model.

63. Orrell et al. 2001. Model convergence is not in itself sufficient for prediction accuracy, since the models could be converging to the wrong thing, but if they have not converged, it is an obvious indicator of a problem.

64. Ibid. See also Matthews 2001. There are probably particular situations where the climate system is sensitive to initial conditions but the net overall effect is small. See also Wolfram 2002, p. 1177. The effect of the data-assimilation procedure is to reduce the contribution of both the apparent forecast errors and the initial-condition errors to the drift calculation (Orrell 2005c).

65. Pearson 1905.

66. This is from an e-mail dated 24 Nov. 2000.

67. Errico et al. 2002.

68. Lorenz 1982; Simmons et al. 1995.

69. Orrell 2002. The effect is like an accounting trick in finance, in which apparently rapid growth is conjured up by transferring money from other accounts. Wolfram also mentions the problem: "Attempts are sometimes made to detect sensitive dependence directly by watching

whether a system can do different things after it appears to return to almost exactly the same state. But the problem is that it is hard to be sure that the system really is in the same state—and that there are not all sorts of large differences that do not happen to have been observed." Wolfram 2002, notes to p. 972.

70. Neumann and Flohn 1987. Quoted in Sherden 1998, p. 18.

71. The technique is known as 4DVAR. See Cohn 1997. Orrell 2005b discusses the effect on forecast errors. See also Houtekamer et al. 2005.

72. Which is why, prior to 2000 when I was writing my thesis, it was viewed as uncharted territory. Papers that appeared to show that model error is small included Downton and Bell 1988, Richardson 1997, and Harrison et al. 1999. These showed that on certain occasions, different models gave quite similar forecasts. But this does not imply that models are right, only that they are wrong in similar ways. More recent papers claiming an enhanced role for chaos are based on lagged forecasts (Simmons and Hollingsworth 2002; Hamill et al. 2006), but this method makes the model appear more chaotic than it really is. A team from the Meteorological Service of Canada performed data assimilation experiments which accounted for model error: "Our results seem to support the paradigm, proposed by Orrell et al. (2001), in which model error is the main error source for the first few days of a forecast." Houtekamer et al. 2005.

73. Kosko 1993, pp. 41–42.

74. Wolfram 2002, notes to p. 971.

75. Ibid., p. 381.

76. As science writer Nigel Calder put it, "The butterfly served as a scapegoat for 40 years." Calder 2003, p. 137.

77. At higher resolutions, "errors in the physical parameterizations and previously negligible effects both become increasingly important." Roebber et al., 2004.

5 ► IT'S IN THE GENES
PREDICTING OUR HEALTH

1. Sequence obtained from www.ensembl.org.

2. Quoted in Keller 2000, p. 84.

3. Galton 2001.

4. Galton 1871.

5. Galton 1869.

6. Galton 1898.

7. Galton 1876. Gillham 2001, p. 211.

8. Galton 1886. Only a distribution is given, so I randomly generated individual samples that are consistent with the original data.

9. Galton 1888. For the correlation to equal the slope of the line, the standard deviation of the scatter in the two data sets should be the same, so a small correction is required here.

10. Human bones from the Classic Maya period, for example, show a divergence in height between the poor and wealthy classes, caused by increasing inequality in the society. Wright 2004, p. 100.

11. Galton 1878. Quoted from Gillham 2001, p. 217.

12. Galton 1891.

13. Galton 1865.

14. Quoted from Gillham, p. 207.

15. Plato's *Republic*, book 5, line 1929. Written 360 B.C. Translated by Benjamin Jowett. Available online at http://classics.mit.edu/.

16. Ibid., line 1961.

17. Ibid., line 1959.

18. Churchill predicted that "the breeding of human beings and the shaping of human nature" would be common by 1982. Churchill 1932. In Alberta, almost 3000 forced sterilizations were carried out between 1928 and 1972, under the Sexual Sterilization Act. Since then, the province has made over $800 million in compensation payments. Van Kampen 2005, p. 41.

19. Quoted from Ridley 2003, p. 185.

20. In plants, the egg cells are fertilized by pollen. Pea plants are self-fertilizing, so to control this, Mendel occasionally "castrated" his plants by cutting off the pollen-producing stamens.

21. Mendel's theory was so little known that it was independently redis-covered in 1900 by three separate people—Hugo de Vries, Carl Correns, and Erich von Tschermak—none of whom was aware of his work.

22. Morgan et al. 1915.

23. Matthews 2000. Tjio and Levan (1956) were the investigators who found that the total number of human chromosomes was forty-six. Painter had used the method of sectioning cells. During sectioning, chromosomes can be cut in half and then identified on an adjacent section as a separate chromosome.

24. Schrödinger 1944.
25. Watson and Crick 1953.
26. Maddox 2002.
27. Nirenberg 1963. The correspondence between the genetic code and the resulting sequence of amino acids in the protein is similar to that between the digital representation of letters in a computer and the resulting text. Computers work in 0s and 1s, or base 2. In genes, each letter is specified by a three-digit string in base 4 (denoted A, C, G, T rather than 0, 1, 2, 3), which can represent up to sixty-four different amino acids. Since there are only twenty such acids, there is some redundancy in the genetic code—more than one string can code for the same thing.
28. Shapiro 1999.
29. Dawkins 1976.
30. As the biotech CEO Michael West put it, "The dream of biologists is to have the sequence of DNA, the programming code of life, and to be able to edit it the way you can a document on a word processor." Quoted in McKibben 2003, p. 12.
31. The sequence was complete in 1995, but publication did not occur until the next year. The yeast genome was divided up among about a hundred laboratories around the world.
32. Blair's comment is strange, since during the 1997 campaign, when asked his chances of victory, he replied: "I have never been in the business of making predictions, I am not now, and I never will be in that business." Ormerod 2000, p. 79.
33. The International Human Genome Sequencing Consortium 2001.
34. Hammock and Young 2005. See also Mattick 2005.
35. In fact, the difference between apes and humans may have less to do with genes, per se, than with the rate at which our bodies develop. Adult humans resemble juvenile or even fetal apes. Our slower pace of development gives our brains greater size, more plasticity, and the ability to learn. Gould 1977, p. 365.
36. Weston and Hood 2004.
37. Van Kampen 2005, p. 85.
38. Van Kampen 2005, pp. 208–17. Myriad offers tests for a number of conditions, including hereditary melanoma: see www.myriad. com. They were assisted in their search for the BRCA genes by the detailed genealogical tables maintained by local Mormon families. In the Canadian province of British Columbia, the patent is effectively ignored. See http://www.bccancer.bc.ca.

39. Gonzalez et al. 2005.

40. Eto et al. 1989. *APOE* is also associated with the development of late-onset Alzheimer's disease.

41. The International HapMap Consortium is searching for sets of genes that tend to vary together in the human population, and may be related to the development of certain diseases. For a discussion, see Goldstein and Cavalleri 2005.

42. Willich et al. 2006.

43. A famous study of British civil servants showed that the strongest predictor for heart disease was job status, apparently because low-status workers had less control over their jobs and experienced more stress. Marmot et al. 1991.

44. For a study of the relationship between longevity and social networks, see Giles et al. 2005.

45. Alter and Eny 2005.

46. Kollar and Fisher 1980.

47. Yamamoto 1985; Glaser et al. 1987.

48. Manning 2002.

49. Ackerman 2001, p. 111.

50. Not quite that simple for people with an "intersex" condition. See Dennis 2004.

51. The Committee for the Prevention of Jewish Genetic Disease in the U.S. now routinely checks schoolchildren's blood for the genetic mutations that cause Mendelian diseases like Tay-Sachs (which destroys the nervous system and usually kills by the age of five). The dangerous form of this gene is nine times more prevalent in Ashkenazi Jews than in the general population. If a child is born with two copies, he develops the disease, so carriers of the gene want to avoid marrying other carriers. See Van Kampen 2005 for a discussion of population genetics programs.

52. Moore 2002, p. 208. See also p. 3.

53. Part of the problem is that in a large number of studies with different groups of people, only those that show a positive correlation get reported. If there are enough studies, at least one is bound to show a positive result. See also Haga et al. 2003.

54. It is important to separate inheritance from genetics. The former often includes many socio-economic factors—we usually inherit a family and social structure as well as genes. An embryo's development is also influenced by non-genetic factors, such as the environment in the womb, as discussed in the text. Finally, as we will discuss later, it is possible for

traits to be the result of certain genes, without being able to compute that trait from a knowledge of the DNA alone. Note that some DNA, known as mitochondrial DNA, is inherited from the mother, while in men the Y-chromosome is inherited from the father. This means that certain traits may be inherited from the mother, others from the father. For example, the mitochondrial DNA seems to be associated with aging. See Martin and Loeb 2004.

55. See Hubbard and Wald 1999.

56. See http://mededu.med.uottawa.ca/oisb/media/pic_news_article.jpg. One such program is the Virtual Cell project at Harvard: http://vcp.med.harvard.edu/research.html. See also Gibbs 2001.

57. The sensitivity of DNA to its immediate cellular environment is apparent in cloning experiments, such as those that produced the ewe Dolly in 1996 and are now done on a fairly routine basis with mice and other animals. In these experiments, a nucleus from the animal to be cloned is injected into an egg cell whose own nucleus has been removed. The resulting cell, placed in the womb of a third animal, therefore contains both the DNA and a large number of proteins. The variability in these proteins, which control among other things the DNA-reading process, is what makes cloning difficult. Dolly, for example, was the single survivor of 277 cloned embryos. It also means that animals cloned from the same DNA are not perfect copies of each other. The closest thing to this are identical twins, who split after conception from the same fertilized egg cell. Owing to small random differences in their development, even they are not really identical.

58. Heath et al. 2003.

59. The figure was obtained using the open source program Cytoscape, with the sample yeast data. See http://www.cytoscape.org.

60. Ramsey et al. 2005 presents one computational tool. As discussed in that paper, it is also possible to obtain a quick estimate of the stochastic variability without doing the detailed calculations.

61. A result is that more than one set of values for the parameters can give the same model output. This problem is termed nonuniqueness or underdetermination. See Oreskes et al. 1994.

62. Lewontin 2001, p. 128. See also Edelman 1987.

63. Saunders et al. 1998.

64. Zhang et al. 2005.

65. The Institute for Systems Biology was set up by the biologist and inventor Leroy Hood, whose company, Applied Biosystems, supplied

the "fluorescent" sequencing technology that helped make the Human Genome Project feasible. Information about the ISB is available at http://www.systemsbiology.org.

66. Over 1,500 papers have been written on the galactose network, dating back to the 1950s. See, for example, Verma et al. 2003, Acar et al. 2005. Our model is described in De Atauri et al. 2004. Pedro de Atauri is a biologist, Stephen Ramsey was the astrophysicist, and I was the resident mathematician. The model was run on a forty-six-node Intel dual-processor cluster computer (IBM). The theoretical analysis of the feedback loops is given in Orrell et al. 2006. The stochastic noise generated by a biological system, which results in the spread of responses, can be estimated using the same drift techniques used for model error in weather forecasting. The stochastic perturbations play the same role as the random atmospheric effects that create weather-prediction errors. The estimation technique allows us to understand how noise is controlled. Details are in Orrell and Bolouri 2004; Orrell et al. 2005.

67. The laboratory experiments were performed by J. J. Smith, M. Marelli, and T. W. Petersen in J. Aitchison's group at the ISB. See Ramsey et al. 2006.

68. The model reactions were only parameterizations of complex processes within the cell. The parameters were underdetermined by the experimental data, but at the same time, they had to be carefully balanced to make the model work. Small random perturbations to parameters would upset this balance, even if the perturbations were far smaller than the uncertainty in the parameters themselves. The sensitivity is a natural result of the combination of feedback loops. Another example is the Daisyworld system in Appendix III, which is robust to external perturbations but sensitive to internal modelling errors. It follows from these that a model of a biological or an ecological system doesn't always need to be robust against random changes in each of its kinetic parameters (at least in terms of exact "point" predictions), and conversely that parameter sensitivity in the model does not necessarily imply sensitivity in the true system. Of course, in nature individual cells often vary significantly, but still manage to function (Morohashi et al. 2002). Similarly, different versions of Daisyworld could be constructed, each of which is homeostatic, but also sensitive to changes in the parameters.

69. As Antoine Danchin wrote, "With the evolution of ever more complex animals, numerous regulatory mechanisms have appeared.

Gradually they have combined to ensure the stability of the environment surrounding the cells within the organism, rather than leaving them subject to major environmental changes. These protection mechanisms seem to be necessary because the enzymes that catalyze the cells' chemical reactions are highly sensitive to their physicochemical environment. This sensitivity is an effective selection procedure that explains how the organism as a whole can normally maintain the local environment of its organs and cells within narrow limits." Danchin 2002, p. 312. The models' sensitivity to parameters is like the mirror image of the organism's sensitivity to its external environment—the more sensitive and fine-tuned the control system, the more easily a model of it is disturbed by internal model error. The human eye exists in many different variations and is in many ways robust, but this does not mean you would trust a doctor to randomly alter some aspect of its function in your own body. In each person a delicate balance is maintained, in what Heraclitus described as "a backward-turning adjustment" or "a harmony of opposite tensions" (depending on translation, see note below).

70. Heraclitus, fragment 51. An alternative translation reads: "by being at variance it agrees with itself, a backward-turning adjustment like that of the bow or the lyre." Guthrie 1968, p. 439. Heraclitus rejected the Pythagorean division of phenomena into two classes, one good, the other evil, and argued instead that "Good and evil are one." Cornford 1969, p. 84.

71. Danchin 2002, p. 109.

72. As Evelyn Fox Keller wrote, the gene "is itself part and parcel of processes defined and brought into existence by the action of a complex self-regulating dynamical system in which, and for which, the inherited DNA provides the crucial and absolutely indispensable raw material, but no more than that." Keller 2000, p. 71.

73. Slonim et al. 2000. One problem is that cancer cells are constantly mutating, so they become resistant to chemotherapies. Statistical models can be helpful in understanding this process.

74. One example of this approach is Genentech's drug Herceptin, developed to target the 25 percent or so of breast cancer cases where the HER2 gene is over-expressed in the tumour. Individual patients also vary greatly in their response to other compounds, such as antipsychotic drugs, and this may be correlated to genetic differences. Goldstein and Cavalleri 2005. As another example, the enzyme denoted EGFR, which

stimulates cell division, is over-expressed in about 80 percent of lung cancers. Drugs exist to block its activity, but their effectiveness depends strongly on the patient, and is highest in the 10 percent of patients who have a particular form of EGFR. Anonymous 2002, 2004.

75. For example, Sciona in Boulder, Colorado, will sell you a "DNA, diet and lifestyle assessment test" which assesses your health risks based on a questionnaire and a DNA sample from a cheek swab (http://www.sciona.com/). Their test was withdrawn from stores in the U.K. in 2002 after Genewatch UK protested that it was misleading (www.genewatch.org). Another company working in this area to develop tests is Craig Venter's Celera Genomics. See also the Center of Excellence for Nutritional Genomics, University of California at Davis, http://nutrigenomics.ucdavis.edu/. The potential for new medicines is a subject of debate. Sir David Weatherall of the Royal Society says "Personalised medicines show promise but they have undoubtedly been over-hyped" (Anonymous 2005g). Cancer treatment is one area that has so far shown real benefits.

76. Quoted in Horgan 1996, p. 125. Gould was diagnosed with abdominal mesothelioma, a rare and serious form of cancer, in July 1982. Median mortality was eight months after diagnosis. He died twenty years later, of another form of cancer. He is the author of many popular books, including *The Structure of Evolutionary Theory* (2002). Another example of how biotechnology can be used to fight cancer is that new diagnostic tests, based on a blood sample, can detect some forms of mesothelioma at an early stage (Cullen 2005).

77. Quoted in Abraham 2005. *The Economist* magazine, for example, predicts that we will soon have "knowledge about how genes interact with each other to generate different personal characteristics. Eventually, it might be possible to engineer a true 'designer baby'—one whose adult looks, and possibly mental characteristics, were chosen by its parents . . ." (Anonymous 2000). As the science writer John Horgan put it, "Designer-baby predictions . . . are laughable, given the track record of behavioral genetics." Horgan 2004.

78. Quoted in Koestler 1968, p. 245.

79. Haga et al. 2003.

80. Quoted in Lewontin 2001, p. 207.

81. Biological reductionism is, of course, still around. In the early 1990s, the U.S. government launched the Federal Violence Initiative, which aimed to detect potentially violent people by testing levels

of the neurotransmitter serotonin. People with low serotonin are somewhat more likely to show violent or impulsive behaviour, and somewhat more likely to die from accidents, murder, or suicide. Drugs such as Prozac are believed to work by affecting serotonin levels in the brain. But serotonin levels are also affected by one's perceived societal status: a decrease in rank tends to lower them. It would make as much sense to stamp out crime by arresting anyone who looks a bit down. The initiative was fortunately squashed in 1992, after its lead organizer made some remarks comparing youths in America's ghettos with monkeys in the jungle. Moore 2002, p. 226.

82. Quoted in Bernstein 1998, p. 216.

6 ► BULLS AND BEARS
PREDICTING THE ECONOMY

1. Mackay 1852.
2. Jonathan Swift, *The South Sea Project,* 1721.
3. Freese 2003, p. 40.
4. Quoted in Bernstein 1998, p. 124.
5. Ibid., p. 141.
6. Ibid., p. 160. Quetelet demonstrated the use of the normal distribution to fit human data by using published data on the chest measurements of 5,738 Scottish soldiers.
7. Orléan 2001.
8. Any such valuation of environmental services will of course be highly subjective and political, but it's better than valuing them at zero, which has led, for example, to the loss of rainforest to agricultural production. In the Brazilian Amazon, about 25,000 square kilometres is lost per year. Nearly half of that occurs in Mato Grosso state, where the world's largest soy farms are located. Anonymous 2005; Anonymous 2005a.
9. I used to pass Jeremy Bentham in the hallway from time to time on my way to teach classes at University College London. He died in 1832, but one of the terms of his will was that his mummified body would be displayed in a glass case, his head replaced with a wax copy. Perhaps he saw it as a way of maximizing his own utility, even after his death.
10. Jevons 1888. See also McCusker 2004, Nicholls 1998. For weather forecasting based on sunspots, see http://www.weatheraction.com/.
11. Galilei 1967, p. 59.

12. Smith 1776.

13. McMurdy 2004.

14. Bachelier 1900.

15. Quoted in Bernstein 1998, p. 200.

16. In Cootner 1964. See also Roberts 1959.

17. Cowles 1933.

18. Fama 1965.

19. As opposed to "alternative" areas of research such as green economics, which are not usually aimed at making specific financial predictions. See, for example, Hawken 1994; Daly and Cobb 1989.

20. A number of websites help you analyze charts; see, for example, www.stockcharts.com.

21. For a discussion of the track record of technical trading with references, see Malkiel 1999, p. 160. The statement applies to average performance.

22. Haugen and Lakonishok 1992.

23. Fama 1965.

24. Williams 1938.

25. Sherden 1998, p. 167.

26. Malkiel 1999, p. 170.

27. Fama 1965.

28. On March 6, 1999, *The Economist* announced on its cover that the world was "drowning in oil." Prices had been falling in a roughly linear fashion for over a year, and had reached ten dollars a barrel. The article ventured a prediction that prices could soon fall as low as five dollars. At about exactly that moment, oil prices changed direction and reached twenty-five dollars by the end of the year (when the magazine followed up with an article entitled "We Woz Wrong"). Anonymous 1999.

29. The Organisation for Economic Co-operation and Development (OECD) notes that until the mid-1990s, many housing markets tended to follow a cycle with an expansion phase lasting around six years, and a contraction phase lasting about five years (maybe Jevons had a point when he linked the economy to the eleven-year sunspot cycle . . .). However, this pattern then seemed to break when low interest rates ushered in a long period of expansion (OECD 2005).

30. Woudenberg 1991. Quoted in Sherden 1998, p. 169.

31. Adapted from Koutsogeorgopoulou 2000, figure A2, p. 45. Results are similar for the individual countries.

32. A 1993 comparison of OECD forecasts with those from the International Monetary Fund and the governments of the USA, Japan,

Germany, France, Italy, and Canada showed no improvement over naïve forecasts. OECD Economic Outlook, June 1993.

33. *The Economist,* July 27, 1991, p. 61. Quoted in Sherden 1998, p. 62. "Of 60 recessions in developed and developing economies during the 1990s, two-thirds remained undetected by consensus forecasts as late as April of the year in which the recessions occurred. In one-quarter of cases, the consensus forecast in October of that year still expected positive growth." Anonymous 2001a.

34. Anonymous 2005b.

35. McCauley 2004, p. 6. For a discussion of models used to predict currency fluctuations, see Cheung et al. 2005.

36. Buchanan 2000, p. 133.

37. McNees and Ries 1992.

38. McCauley 2004, p. 6.

39. Malkiel 1999, p. 242. "Where it counts, the market's behavior conforms to the rational model." Bernstein 1998, p. 296.

40. Aristotle et al. 1981, p. 90.

41. Kenneth Arrow described this discrepancy as an "empirical falsification" of orthodox theory. See Ormerod 2000, p. 16.

42. Liu et al. 1999.

43. See McCauley 2004, p. 7; Ormerod, p. 128 for examples with economic variables.

44. Buchanan 2000, p. 157.

45. See Barabási and Albert 1999; Zipf 1949.

46. The number of connections that proteins have in biological networks also follow a power-law distribution. Barabási and Oltvai 2004. See Buchanan 2000 for a general discussion.

47. Dunphy 2004.

48. See Homer-Dixon 2000, pp. 292–96 for a discussion.

49. Simons 1997.

50. Bollerslev 1986; Engle 1982.

51. Mandelbrot and Hudson 2004, p. 104. See also Loomes 1998, Anonymous 1999a.

52. Singer 2005.

53. Ormerod 2000, p. 72.

54. Iamblichus 1918.

55. See, for example, the above quote from Malkiel 1999, as well as Bernstein 1998, p. 296.

56. "Double-think" is Arthur Koestler's expression. Koestler 1968.

57. Quoted is the Nobel laureate economist Merton Miller, from Malkiel 1999, p. 273.
58. Soros 2000, p. 27. Dangers of an unfettered market: George Soros interview, 20 Sept., 2002. "Wall $treet Week with FORTUNE," PBS.
59. Bass 1999.
60. See, for example, Lux and Marchesi 1999.
61. See McCauley 2004 for a discussion. Companies using agent-based simulations for specific business problems include NuTech Solutions (http://www.biosgroup.com/) and Volterra Consulting (http://www.volterra.co.uk/).
62. See Ormerod 2000, p. 200, for an example of a conceptual model that incorporates internal prediction errors.
63. Quoted in Bernstein 1998, p. 203.
64. Soros 2000.

7 ► THE BIG PICTURE
HOW WEATHER, HEALTH, AND WEALTH ARE RELATED

1. See Wright 2004.
2. Ehrlich 2000, p. 245; Kirch 1984; Diamond 2005.
3. Maunder 1997.
4. Asimov 1951.
5. Popper 1957, p. 4.
6. Meadows et al. 1972, p. 23.
7. Ibid., p. 58.
8. See *The Population Bomb* by Paul Ehrlich (1968). He overestimated the rate of U.S. population growth, which slowed after the introduction of the contraceptive pill. The popularity of his book may also have played a contraceptive role.
9. A number of U.S. politicians seem to believe in Armageddon. For example, James Watt, who was President Reagan's first secretary of the interior (not the inventor of the flyball governor), testified to the U.S. Congress that protecting natural resources was unimportant since "after the last tree is felled, Christ will come back." Moyers 2004.
10. According to the Serbian mathematician Milutin Milankovitch, whose 1924 theory appears consistent with recent ice age estimates.
11. Plato 1961.
12. Ruddiman 2005.

13. Ehrlich and Ehrlich 2004, p. 201.

14. Ibid., p. 21.

15. Ibid., p. 42.

16. Schnur 2002.

17. Kennedy 2004.

18. Nepstad et al. 1999.

19. It is impossible to accurately measure the rate of species extinction, but scientists generally agree that it has increased dramatically since our arrival. According to Martin Rees, it has increased from about one species in a million per year to one in a thousand. Rees 2003, p. 101.

20. Stocks of smallpox are maintained at the Centers for Disease Control, Atlanta, and the Vector Laboratory in Moscow.

21. 1.01^{2000} is slightly more than 439 million.

22. For example, Margolis 2000, p. 83: "The very contention that a relatively puny species . . . has damaged, and without really trying to, something as vast as an entire planet will strike many as an especially audacious example of the arrogance of the present." This refers to "the belief of every generation that it alone has been chosen to live in a 'special' time. . . . The first cousin of plain egotism."

23. Crichton 2004, p. 570. See also Allen 2005. Personally, I agree more with the introduction to Crichton's 2002 thriller *Prey,* written before he got into climate research: "The fact that the biosphere responds unpredictably to our actions is not an argument for inaction. It is, however, a powerful argument for caution, and for adopting a tentative attitude toward all we believe, and all we do."

24. This price mechanism is not straightforward, because it only works if the producers are aware of the scarcity. Norgaard 1990.

25. Huber and Mills 2005.

26. Skinner 1973, pp. 10–12.

27. Stoto 1983.

28. See Wright 2004; Diamond 2005.

29. Population Reference Bureau 2000. See www.prb.org.

30. Ehrlich and Ehrlich 2004, p. 185.

31. World Wildlife Fund, www.panda.org.

32. Millennium Ecosystem Assessment, 2005. Statement of the MA Board. See www.maweb.org.

33. Rees 2003.

34. Deacon 1968.

35. Nordhaus 1994a.

36. Suzuki 2005. Suzuki's website is http://www.davidsuzuki.org.

37. Arrhenius 1896. The warming effect of greenhouse gases was first noted by the British physicist John Tyndall in 1859.

38. Actually, Arrhenius thought that global warming would bring benefits, at least for Sweden. Extrapolating from the rate of production at that time, he estimated it would take 3,000 years for atmospheric carbon dioxide to double.

39. Kerr 2004b.

40. Ibid. The word "canonical" is used in this way in Kerr 2005.

41. Jasanoff and Wynne 1998, p. 70. Quoted in Rayner 2000, p. 275.

42. Quoted from Stephen Schneider's website in August 2005 (http://stephenschneider.stanford.edu/).

43. Whitfield 2003.

44. Over millions of years of the planet's history, trillions of tons of carbon were removed in this way, buried in sediments and slowly transformed into fossil fuels such as coal, oil, and natural gas.

45. It is hard to assess what the net effect of this activity is on the biosphere. Some is quickly reabsorbed by plants, to whom more CO_2 means more food, so they grow at a faster rate. An acre of healthy, young forest can remove carbon from the atmosphere at a rate of up to a ton per year, and five acres can compensate for the carbon emissions of an average North American. The amount of carbon dioxide that can be absorbed by greenery or by the oceans is, however, limited. Plants can't grow forever, and when they die and decay, or burn in forest fires, they release their stored carbon. The ability of oceans to absorb CO_2 is similarly constrained by ocean chemistry, and will eventually reach saturation.

46. Wohlforth 2004, p. 147.

47. The IPCC writes: "Probably the greatest uncertainty in future projections of climate arises from clouds and their interactions with radiation." IPCC 2001a.

48. Errico et al. 2002.

49. This question has been raised by skeptic lobbying organizations such as Americans for Balanced Energy Choices, who write on their website: "Predicting *weather* conditions a day or two in advance is hard enough . . . so just imagine how hard it is to forecast what our *climate* will be 75 to 100 years in the future." The point seems naïve, but it's not. Short-term predictability and long-term predictability are linked because model errors affect each—unless you believe that all error is caused by chaos.

For simple examples, see Orrell 2003, and the discussion of the Lorenz system in the Appendices.

50. Quoted in Zöllner and Nathan 2003, p. 76.

51. In 2004, the National Center for Atmospheric Research (NCAR) model indicated that the amount of low-level cloud cover will increase, while the GFDL model said it should decrease. As a result, the GFDL predicts an albedo change that is a factor of three greater. Kerr 2004b.

52. In Charles Wohlforth's book *The Whale and the Supercomputer*, he describes a NASA experiment in which satellite measurements of floating Arctic ice were compared over different scales, with observations from an aircraft flying as high as 6,000 metres, an unmanned Aerosonde aircraft at low altitude, a team of scientists standing on the ice, and a Native elder. From outer space, a 200 km^2 patch consists of a few pixels in the satellite image. The observer in the airplane can make out large patches of snow, ice, and water, and the Aerosonde can pick out distinct features on the surface (such as small variations in height). To the people on the ground, though, the ice seemed "easier to take in as pure chaos. . . . At every scale there were textures and patterns being broken and superseded." One of the scientists is depressed at the difference between their models and reality. The Iñupiaq elder thinks that the movement of ice is too complex to model or predict, and what can be known just comes down to common sense: "They use science to prove things we already know." Wohlforth 2004, p. 90.

53. "Flux adjustments are non-physical in that they cannot be related to any physical process in the climate system and do not a priori conserve heat and water across the atmosphere-ocean interface. . . . The approach inherently disguises sources of systematic error in the models, and may distort their sensitivity to changed radiative forcing." IPCC 2001a. An alternative is to tune the parameters to give a stable model climate.

54. Stainforth et al. 2005. The paper notes: "The range of sensitivities across different versions of the same model is more than twice that found in the GCMs used in the IPCC Third Assessment Report." Over 2,000 simulations were performed. The selected runs were compatible with climate observations, so in this sense the perturbed parameter values were not unrealistic. Only six parameters were perturbed, which affected the representation of clouds and precipitation. These were "the threshold of relative humidity for cloud formation, the cloud-to-rain conversion threshold, the cloud-to-rain conversion rate, the ice fall speed, the cloud fraction at saturation and the convection entrainment

rate coefficient." The paper points out that the range of perturbations may be too low because "experts are known to underestimate uncertainty even in straightforward elicitation exercises." Also, "even the physical interpretation of many of these parameters is ambiguous." There are, of course, many other parameters in a GCM that could be perturbed. The experiment was carried out not on a supercomputer but by using the idle processing capacity of tens of thousands of computers volunteered by the public. See www.climateprediction.net.

55. Of course, this is very useful information, and it was a good experiment. It has been mistakenly interpreted by some authors to imply that the mean and variance of the results says something about the climate. The mean only reflects the original choice of model around which perturbations were made, and the variance only reflects the model's sensitivity to parameterization, for those particular (and rather small) perturbations. In other words, the results say a lot about the model, but not much about the climate system itself.

56. Allen et al. 2002. See also Smith 2000.

57. Wohlforth 2004, p. 169.

58. Kerr 2005.

59. Rayner 2000. Koestler described Ptolemy's final version of the Greek Circle Model as "a monumental and depressing tapestry, the product of tired philosophy and decadent science" (Koestler 1968, p. 69). Have we come full circle? The former Secretary-General of the UN World Meteorological Organization, Aksel Wiin-Nielsen, wrote in 1996, "The most important explanation as to why so much extensive theoretical work in the development of climate models has been done during the last ten years is that the development of models sustains funding and secures jobs at research institutions." Quoted in Lomborg 2001, p. 37.

60. Another example is the yeast galactose network discussed in Chapter 5. The model accurately predicted that removing regulatory feedback loops would make the yeast behave in a more erratic fashion, but it was impossible to be sure of this until the experiment was performed. It would have been very hard to defend the model against a skeptic who didn't believe the prediction, or to assign probabilities to different outcomes, because we knew that the model was only a coarse approximation of the real thing. We had no idea if there were other feedback loops that could compensate for those that had been removed, or if the mutant yeast would even function as an organism. Pretending to make

a probabilistic forecast for a single yeast cell would have been inappropriate, to say the least.

61. For evidence based on measurements of the ocean's heat content, see Hansen et al. 2005. For information about the state of sea ice, see the National Snow and Ice Data Center (http://nsidc.org). See also Kolbert 2005 for a general discussion.

62. Parmesan and Galbraith 2004.

63. The climate has always been variable, and life on this planet has somehow survived numerous ice ages, along with smaller effects (such as volcanoes or comets) that cool the climate by throwing dust into the atmosphere. In the past, species could often migrate with the climate, moving to cooler locations as it warmed and vice versa. Some went extinct, as species do, but a robust ecosystem is flexible and can adapt well to external perturbations. Such migration is now far more difficult, because of obstacles such as roads, farms, urban conurbations, and so on. Winged species have an advantage, but laggards arriving by foot will take longer, and some trees and plants, which can move only by dispersing seeds, will find it hard to survive a rapid climate perturbation. Ocean-dwelling species will have their own challenges. Schneider and Root 2001; Tol and Dowlatabadi 2002; Byers 2005; Barnett and Adger 2003.

64. What matters in climate is not so much the average warming over the globe but the local climate in particular regions. If, for example, some areas experienced extreme heating while others were cooled, or some became very wet while others suffered from drought, then the small average change worldwide would mask significant local effects. It is even harder for climate models to pick up these subtle local variations than it is to detect the broader warming trend. IPCC 2001. For example, while the NCAR and GFDL models agree quite well on average warming, the first predicts a wetter United States, the second a drier.

65. See http://www.seaaroundus.org/ for information on the state of fisheries.

66. Baker et al. 2005, Watson 2005. The average historical return is about 6.5 percent. The equations used are simple and make all sorts of assumptions about productivity growth, population growth, and the world at large. Results therefore depend on the judgment and biases of the forecaster.

67. IPCC 2000.

68. IPCC 2001.

69. The latter adjusts wealth according to domestic purchasing power. Anonymous 2005d.

70. Peter Schwartz and Doug Randall, "An Abrupt Climate Change Scenario and Its Implications for United States National Security," October 2003. Quoted in Byers 2005.

71. Nordhaus 1994. See Rotmans and Dowlatabadi 1998 for a discussion of integrated assessment models. Information about RICE and DICE is available at the homepage of William Nordhaus (http://www.econ.yale. edu/~nordhaus/homepage/homepage.htm).

72. This from *The Skeptical Environmentalist* by Bjorn Lomborg, who writes that the model "gives the same qualitative conclusions as all other integrated assessment models." Lomborg 2001, pp. 306, 310. Costs are adjusted for the year 2000.

73. Quoted in Bernstein 1998, p. 203.

74. Ziegler 1965.

75. Osterholm 2005.

76. Matthews and Fraser 2000.

77. The name H5N1 refers to proteins on the surface: hemagglutinin type 5 and neuraminidase type 1. The 1918 virus has been artificially reconstructed. See Tumpey et al. 2005.

78. Garrett 2005. The mortality rate for avian flu may be an overestimation, since less serious cases are probably not all reported.

79. Anonymous 2005c.

80. In October 2005, estimates of potential deaths from a bird flu epidemic, should it occur, range from 2 million to 360 million. As the World Health Organization spokesman Dick Thompson said, "One of those numbers will turn out to be right. We're not going to know how lethal the next pandemic is going to be until the pandemic begins." Anonymous 2005e.

81. Watts 2003, p. 169. The track record of such models is not very good. An example was a model used to simulate the 2001 outbreak in U.K. livestock of the highly infectious foot and mouth disease. Despite being highly simplified, the model was used as the basis of a decision to cull millions of cows. The action was later judged as far too draconian—one report (Campbell and Lee 2003) called it "carnage by computer."

82. Eubank et al. 2004.

83. As Antoine Danchin said, "It is impossible to predict the future of microbes and parasites." Danchin 2002, p. 274. We can follow the evolution of viruses, and perhaps detect whether they are gaining

attributes that make them more dangerous, but any kind of precise forecasting is impossible.

84. Wichman et al. 1999.

85. Jackson et al. 2001.

86. See http://www.phac-aspc.gc.ca/media/nr-rp/2004/ 2004_gphin-rmispbk_e.html.

87. Longini et al. 2004. One such drug is Tamiflu, which in 2005 was being stockpiled by both governments and individuals, sending the shares of its manufacturer, Roche, up by 40 percent in less than a year. Foley 2005. As Christina Pearson, spokeswoman for the U.S. Department of Health and Human Services, pointed out: "We don't know right now what the next pandemic strain will be. It's uncertain if it's going to be H5N1 (bird flu). It's uncertain whether Tamiflu, or other things—how effective they will be against that strain. Part of our strategy will be to stockpile antivirals. To put it bluntly, you don't want to put all your eggs in one basket." Greene 2005.

88. Osterholm 2005; Ravensbergen 2004. See also the European Centre for Disease Prevention and Control (ECDC) website at http://www. ecdc.eu.int/.

89. Plender 2005.

90. In October 2005 the U.S. government announced a pandemic prevention plan costed at $7.1 billion. See http://pandemicflu.gov/. A rough estimate from the WHO was also in the billions of dollars (Jack et al. 2005). The World Bank estimates the global cost of a pandemic to be about $800 billion a year.

91. Anonymous 2005c.

8 ► BACK TO THE DRAWING BOARD
FIGURING OUT WHERE WE WENT WRONG

1. The United States Geological Survey website is located at http://pubs.usgs.gov/fs/2000/fs036–00/.

2. Vedantam 2004.

3. See Ridley 2003, p. 60; Ehrlich 2000, p. 111.

4. Klin et al. 2002.

5. Bazell 2005.

6. Quoted in Baron-Cohen 2003, p. 169.

7. Baron-Cohen, together with Ioan James from Oxford University, proposed Newton and Einstein (Muir 2003).

8. A lack of interest in communication is as common in science as in people with Asperger's. In his book *Voltaire's Bastards,* John Ralston Saul wrote, "When faced by questioning from non-experts, the scientist invariably retreats behind veils of complication and specialization. Of course it is complicated. But there is no other profession in which the sense of obligation to convert the inner dialect into the language of man is so absolutely absent" (Saul 1992, p. 79). Talking directly to the media or general public is strongly discouraged. The climate scientist Stephen Schneider wrote: "The unwritten rules in science decree that recognition is supposed to be based on years of careful work backed up by scores of publications appearing in the most strictly peer-reviewed scientific articles dealing with narrowly defined topics . . . not clever phrases that capture the public's—or worse, the media's—attention" (Schneider 1989, p. 201). There are many excellent popular science books, including those cited in the notes, but science is still a foreign country to most people. Indeed, one of the highest-selling books on the history of science, Bill Bryson's *A Short History of Nearly Everything,* was written by a travel writer, as if science were a distant and exotic land, a kind of Patagonia of the mind.

9. Student petition of *autisme-économie.* See http://www.paecon.net/.

10. Ridley 2003, p. 141.

11. Lewontin 1991, p. 51.

12. Knight 1921.

13. It is frequently aimed against environmentalists, despite their often high degree of scientific literacy. See, for example, Taverne 2005.

14. Ingram 2005, p. 211. Patients with damage to the right hemisphere also find it harder to get jokes, which often depends on making a sudden shift in context (Coulson and Williams 2005). In general, the two hemispheres work in concert, so they are as closely entwined as nature and nurture. It is only possible to tease apart their roles in rather contrived situations. To say that a particular individual is "left-brained" or "right-brained" is only a loose statement.

15. Keller 1985, p. 69. Pythagoras allowed women into his group, but as Guthrie points out the female was associated in the column of opposites "with evil, darkness and the unlimited." This may have influenced Plato to describe women as originating from morally defective

souls in *Timaeus* (Guthrie 1978, p. 307). Aristotle excluded them from the Academy, and women have been underrepresented in science ever since—which has undoubtedly affected its course. See Wertheim 1995.

16. Wertheim 1995, p. 100.

17. Griffiths 2004, p. 167.

18. Nelson 1996. Male and female here refer to abstract cultural properties. This does not imply that women are no good at math.

19. Quoted in Koestler 1968, p. 197. Even the science of dynamics seems less about motion than about stopping it so it can be studied frame by frame.

20. Ibid., p. 397.

21. Ibid., p. 329.

22. These fields, including also nonlinear dynamics, fractals, and fuzzy logic, have shifted the emphasis to the right-hand "evil" column (Orrell 2006). Some scientists still dismiss the study of holistic phenomena as a complete non-starter because, it is argued, scientific deduction always has to build from first principles. As Mary Midgley put it, "During much of the twentieth century the very word 'holistic' has served in some scientific circles simply as a term of abuse" (Midgley 2000, p. 14). But holistic phenomena are not the product of some kind of fuzzy philosophy; they are simply a fact of life. Complex systems have emergent properties that cannot be deduced from simple laws. To ignore them is like tying your left hand behind your back: an unnecessary handicap.

23. Midgley 1985, p. 25.

24. For example, one group wrote, "Lack of any conceivable objective verification/falsification procedure has led some commentators to conclude that forecasts of anthropogenic climate change are fundamentally subjective. . . . Worse still, assessment of forecasts of anthropogenic climate change degenerates all too easily into a dissection of the prior beliefs and motivations of the forecasters, 'placing climate forecasts in their sociological context.' As die-hard old-fashioned realists, we firmly reject such New Age inclusivity. . . . To resign ourselves to any other position on an issue as contentious as climate change is to risk diverting attention from the science itself to the possible motivations of the experts or modelling communities on which current scientific opinion rests." Allen et al. 2002. See also Schneider 2002.

25. Koestler, 1968, p. 284.

26. Peterson 1993, p. 41.

27. The topic of validation, a term used very loosely by climate scientists, is discussed in Oreskes et al. 1994: "Even if a model result is consistent with the present and past observational data, there is no guarantee that the model will perform at an equal level when used to predict the future."
28. Keller 1985, p. 141.
29. Ibid., p. 167.
30. Lovelock 1979.
31. Lovelock 1991, p. 22.
32. Quoted in Zöllner and Nathan 2003, p. 76.
33. Lovelock 1991, p. 3; Mann 1991.
34. Lovelock 1991, p. 11. See also Keller 2002.
35. Julie A. Nelson wrote: "The central model of economics views people as individuals, and each individual as self-interested, autonomous, rational, and free to choose among different actions. Logically, the converse of this would be a view of people as linked to others and concerned about their welfare—people who are dependent, emotional, and subject to decisions made by others or influences from the social or natural environment. Not just coincidentally, all the characteristics in the first list have been, in modern Western and English-speaking cultures, associated with stereotypical masculinity, while all those in the latter list are associated with stereotypical femininity." Nelson 1996.
36. Hillman 1975, p. 132; Shamas 2003.
37. For example, what is the point in fighting for more equal distribution of wealth if you believe what James Watson did? "Maybe one of the reasons for this growing inequality of income may in some sense be a reflection of some people being more strong and healthy than others. Some people, no matter how much schooling you give them, will never really be up to what is now considered a necessary degree of effective intelligence" (Duncan 2003). There is a strong political element to these myths and metaphors. The idea that we exist only to further the aims of our selfish genes helped rationalize economic theories based on self-maximizing behaviour and the "no such thing as society" ethos of the Reagan/Thatcher years. Priests used to justify the world order by invoking the word of God; now it is written in our genes. The Czech poet/president Václav Havel prematurely hoped in 1992 that the collapse of the Soviet Union would help end the "cult of objectivity," which was "dominated by the culminating belief, expressed in different

forms, that the world—and Being as such—is a wholly knowable system governed by a finite system of universal laws that man can grasp and rationally direct for his own benefit." Quoted in Horgan 1996, p. 23.

38. Wertheim 1995, p. 29. See also Lerner 1987.

39. Why are we so much more concerned with the rather airy and abstract problem of climate change than with more immediate and equally serious issues such as deforestation, overpopulation, and soil degradation? In November 2000, for example, flooding in the U.K. was blamed by politicians on global warming, rather than the fact that overdevelopment and changes in farm practices had reduced the available area for drainage (Anonymous 2000a). Perhaps it is because it is easier to talk about climate change than fix problems on the ground.

40. Part of the resistance to Gaia theory is that it adopts a holistic perspective, which has traditionally been resisted by scientists. The self-regulating properties of Daisyworld, for example, can be understood only by viewing the system as a whole, rather than from the perspective of either species of daisy in isolation. As the authors of a report from the British think-tank Demos wrote, "That we need a new mental model for our place in the world is increasingly apparent. . . . Gaian thinking can help us to develop a more holistic understanding of ourselves, our organisations, and the needs of our habitat." From John Holden's introduction to Midgley 2000, p. 7.

41. Evolution here refers to a process of growth and change. The planet itself has not been through a process of Darwinian selection, rather it is the end product of the evolution of the species that inhabit it.

42. The metaphor of humanity as a disease has been used often. See, for example, Ehrlich 1968; Worldwatch Institute 1998.

43. Kirby 2004.

44. See the report "Americans Still Not Highly Concerned About Global Warming" at http://poll.gallup.com. About a third of Americans think climate change will be a serious threat during their lifetimes. There are a number of factors that might lead to a sense of detachment about the environment; one is that responsibility is shared by us all and cannot be pinned on a small group. However, science has played a major role in shaping our culture and therefore our attitude towards nature.

45. Meadows et al. 1992, pp. 109–10.

46. The linguist Noam Chomsky wrote: "It is quite possible—overwhelmingly probable, one might guess—that we will always learn more about human life and human personality from novels than from scientific

psychology. The science-forming capacity is only one facet of our mental endowment. We use it where we can but are not restricted to it, fortunately." The same may apply to finding our place in the living biosphere. Chomsky 1988, p. 159. Quoted in Horgan 1996, p. 152.

47. D. T. Suzuki, in the introduction to Herrigel 1953, p. vii.

9 ► CONSULTING THE CRYSTAL BALL
OUR WORLD IN 2100

1. According to the IPCC's "Third Assessment Report," the ice sheet will begin to melt if local temperatures increase by 3°C, which is equivalent to global warming of about 1.5°C.
2. Oppenheimer and Alley 2004.
3. ACIA 2004.
4. Ni 2001.
5. See the Millennium Ecosystem Assessment 2005 for a study of desertification.
6. According to H. R. Kaufmann, general manager of Swiss Re, "Failure to act [on global warming] would leave the insurance industry and its policyholders vulnerable to truly disastrous consequences." *World Watch* 7, Nov./Dec. 1994.
7. Meehl and Tebaldi 2004.
8. The thermohaline circulation is a kind of massive oceanic conveyor belt powered by differences in ocean temperature and salinity. It has the effect of bathing Europe in warm water from the equator. In the 1980s, Suki Manabe and Ron Stouffer showed that the circulation pattern could be cut off by adding fresh water to the Arctic and North Atlantic—which could happen if global warming brings increased precipitation and melting Arctic ice. This would plunge countries such as the United Kingdom into Canadian-style winters. Again, it is not possible to predict exactly what will happen: estimates range from no significant change to a weakening of around 50 percent. IPCC 2001a.
9. Canada's East Coast ecosystem has been so altered by overfishing that cod may never recover. Frank et al. 2005.
10. Hoegh-Guldberg 1999.
11. Leemans and Eickhout 2004.
12. Harvell et al. 2002.
13. Garrett 1994.

14. Preston 2002.

15. This assumes an average annual return of about 6.8 percent, which is not far off the historical rate of return for the U.S. stock market. Baker et al. 2005; Watson 2005.

16. In 1995, the vice-president of the World Bank, Ismail Seageldin, predicted, "If the wars of this century were fought over oil, the wars of the next century will be fought over water." Quoted in Shiva 2002, p. 1.

17. Odds here are about 50 percent for an impact sometime in the century, though it probably won't hit a city. Rees 2003, p. 92.

18. Nanotechnologists stay up at night worrying about this. See Drexler 1986.

19. Sagan and Turco 1990.

20. James Lovelock, for example, supports the use of nuclear power. Lovelock 2006.

21. Michael Crichton predicts that people in 2100 will "have a smaller global population, and enjoy more wilderness than we have today" (Crichton 2004, p. 570). Unless the human race goes off sex, the only way this can happen in one hundred years is if death rates go through the ceiling, as from a war or a pandemic. Perhaps the subject of his next novel . . .

22. Ralph Abraham describes these revolutions in terms of the different types of attractors for a dynamical system. The agricultural revolution corresponds to a shift towards a stable attractor, the Industrial Revolution to a periodic attractor, and the current revolution to a chaotic attractor. Abraham 1994.

23. "If she finds her system getting out of kilter because one element in it is insatiably greedy, she simply ditches that element as she has done so many others before. She is not in the least anthropocentric and has no special interest in intelligence. She is in fact impersonal, impartial Nature—not especially red in tooth and claw, but resolute to remain in general green and alive, and therefore liable to cross the projects of those who are acting so as to turn the green, thriving world into a desert. And if this resistance fails, she herself can no doubt be killed with all her children. No universal fail-safe mechanism protects either her or us." From Midgley 1985, p. 64.

24. From IPCC 2001, table SPM-1.

25. I exaggerate about economists, though a group of them did rank climate change at the bottom of a list of pressing global problems. See Anonymous 2005d, and http://www.copenhagenconsensus.com.

26. Kahn and Wiener 1967; Albright 2002.
27. As one New Orleans blogger wrote, just hours before Hurricane Katrina struck land and flooded his city: "If you stop and think about it for a moment, there's something incredibly humbling about the situation we're in right now, watching and waiting to see where Katrina goes. A week from now, the city of New Orleans—a great, industrialized city in the most powerful nation in the history of the world—might be annihilated, or it might be devastated but not destroyed, or it might be mildly damaged, or it might be perfectly fine. We have absolutely no control over, and a very limited ability to predict, which of these scenarios will occur. We are utterly at nature's mercy." "Humbled by Katrina," posted by Brendan Loy on 26 Aug. 2005 at http://www.brendanloy.com.
28. Abraham 1994, p. 215.

APPENDICES

1. The reason that the average distance between two random points is ⅓ is that two points divide the line segment into three smaller segments; if the points are chosen at random, then the length of each segment will on average be the same. The lengths must add to 1, so the average length of each segment, and therefore the average distance between the points, is ⅓.
2. Just as numbers in the decimal system are represented as their unique decimal expansion in powers of 10, so we can write a unique binary expansion in powers of 2. For example, 142 can be written as $142 = 1 \times 10^2 + 4 \times 10^1 + 2 \times 10^0$ (a number raised to the power 0 is 1), or equivalently as $142 = 1 \times 2^7 + 0 \times 2^6 + 0 \times 2^5 + 0 \times 2^4 + 1 \times 2^3 + 1 \times 2^2 + 1 \times 2^1 + 0 \times 2^0$, which leads to the binary representation 10001110. Fractions are similarly represented using negative exponents.
3. See Wolfram 2002, pp. 149–53.
4. Lorenz 1963.
5. See also Orrell 2003 for other examples.
6. Orrell 2001.
7. This moral is, of course, different from the original "butterfly effect" moral, which implies that forecast error is primarily the result of sensitivity to initial condition.

8. Watson and Lovelock 1983.

9. As Lovelock wrote, "Neither Lynn Margulis nor I have ever proposed that planetary self-regulation is purposeful. . . . Yet we have met persistent, almost dogmatic, criticism that our hypothesis is teleological." Lovelock 1991.

10. See, for example, http://www.acad.carleton.edu/curricular/GEOL/DaveSTELLA/Daisyworld/daisyworld_model.htm or http://gingerbooth.com/courseware/pages/demos.html#daisy.

11. Heraclitus, fragment 54.

12. Hinkle 1996.

Glossary

Amino acid: One of the twenty different molecules that are combined in a linear fashion to form proteins.

Analysis: In meteorology, this word refers to a weather centre's estimate of the state of the atmosphere. It is obtained by compiling observed data from a variety of sources—such as ground stations, weather balloons, commercial airplanes, and satellites—and reconciling it with model predictions from an earlier time. The analysis then forms the initial condition for a new forecast.

Attractor: The trajectories of a dynamical system can be drawn to one of three basic types of attractors. A point, or steady-state, attractor is a single fixed point. In a periodic attractor, trajectories are drawn into a repeating cycle. The third class is the strange attractor, which is characteristic of chaotic systems (see Appendix II). Small changes in parameter values can sometimes cause a sudden bifurcation from one kind of attractor to another. Attractors can be thought of as imposing a kind of constraint on a system's behaviour. They are properties of sets of mathematical equations, not real physical systems, so we can talk about attractors for climate models, but not for the climate itself.

Avian flu: A type of influenza virus that is hosted by birds but may also infect other animals, including humans, where it can lead to acute respiratory distress and pneumonia. At the time of writing, the strain known as H5N1 has been identified by many scientists as the most likely source for the next pandemic. For that to happen, though, the strain will have to genetically mutate so that it can be easily transmitted from person to person without losing its virulence.

BASE: One of the four substances (adenine, cytosine, guanine, and thymine) that make up DNA.

BELL CURVE: See normal distribution.

BIFURCATION: In a dynamical system, a bifurcation occurs when a small change in a parameter causes a qualitative change in the system solution. A jump from a periodic attractor to a chaotic attractor would be an example of a bifurcation.

BUTTERFLY EFFECT: This is the theory that the small atmospheric disturbance caused by a butterfly flapping its wings can cause a storm on the other side of the world. It was inspired by the sensitivity to initial condition of chaotic systems like the Lorenz system (see Appendix II). While the butterfly effect is an attractive idea, it doesn't account for the local damping of small perturbations, and as the science writer Matt Ridley (2003) notes, it probably says less about the atmosphere than it does about the human desire "to preserve linear causality in such systems" (compare *emergence*). The butterfly effect is also used to describe the sensitivity to initial condition of weather models (as opposed to the weather itself), which is a real but rather weak effect.

CELLULAR AUTOMATON: This is a mathematical system consisting of a grid of coloured cells that evolve in discrete time steps according to local rules. The simplest grid is a one-dimensional line (see Appendix I for an example). In two dimensions, the grid can be square, triangular, or hexagonal. Higher-dimensional systems can also be constructed. The cells can be black or white (binary), or a discrete set of colours, or a continuous range. The simplest rules specify the colour of the cell based on the colour of adjacent cells.

CHAOS THEORY: This refers to the study of non-linear dynamical systems characterized by aperiodic behaviour and sensitivity to initial condition. See also *attractor* and *butterfly effect*.

CHROMOSOME: A cell's DNA is divided and tightly folded into separate chromosomes. These earned their name (from the Greek words *chromo*, for "colour," and *somos*, for "body"), well before their function was understood, because they were easily made visible under the microscope by staining.

CLIMATE: This refers to properties of a region's long-term weather. Quantities of interest include average and maximum/minimum rainfall and temperature for a particular time of year.

CLIMATOLOGICAL FORECAST: This is a naïve forecast for a system's future based on long-term statistics. An example would be predicting that the temperature next week will be equal to the average temperature for that time of year.

COMPLEX SYSTEM: There is no shortage of mathematical definitions for this expression. In the context of cellular automata, a complex system is generally one in which the behaviour is neither entirely ordered nor entirely random, but somewhere in between — in other words, interesting. Such systems, though based on simple local rules, have emergent properties (see *emergence*) that cannot be understood in terms of those rules.

CORRELATION: This statistical term defines a relationship between two sets of randomly varying data. A strong correlation implies that the two tend to vary together, while no correlation means that they are completely independent. For example, lung cancer correlates with smoking—that is, those who smoke are more likely to develop the disease. Many correlations are falsely reported. Chance alone ensures that if enough experiments are performed to test for a relationship between two factors, there is a high probability that at least one will show a positive correlation. Of all the experiments, it will be the only one that is deemed interesting and makes it into a scientific report.

DNA (DEOXYRIBONUCLEIC ACID): A long molecule in the form of a double helix, DNA is composed of the four bases (A, C, G, and T) that encode genetic traits.

DRIFT: The model drift is one technique for estimating the component of forecast error owing to both errors in a mathematical model and uncertainty in the observations of the underlying system. To calculate model drift, one performs a number of short forecasts at a sequence of observation times. The drift can be compared with the results from shadow experiments.

DYNAMICAL SYSTEM: This mathematical term refers to a set of equations that specifies the rate of change of a number of variables with time. A solution of a dynamical system is a trajectory starting from a specified initial condition. The equations modelling the fall of a stone in Chapter 2 make up a simple dynamical system. Note that the falling stone itself is not a dynamical system—it is a stone. The mathematical equations are not the same as the underlying reality. See also *attractor*.

EMH (EFFICIENT MARKET HYPOTHESIS): There are different versions of this theory, but an efficient market was originally defined by Fama (1965) as "a market where there are large numbers of rational profit maximizers actively competing, with each trying to predict future market values of individual securities, and where important current information is almost freely available to all participants." The market price for an asset is a "good estimate of its intrinsic value." Any deviations will be small and random, so "actual prices of securities will wander randomly about their intrinsic values." It follows that in an efficient market, no investor can exploit price discrepancies based on fundamental analysis or any other method.

EMERGENCE: A property or behaviour of a complex system that arises from basic laws or principles but cannot be predicted from them. The word was used to describe living systems by philosophers such as Samuel Alexander (1920), who wrote: "Physical and chemical processes of a certain complexity have the quality of life. The new quality life emerges with this constellation of such processes, and therefore life is at once a physico-chemical complex and is not merely physical and chemical. . . . The existence of emergent qualities thus described is something to be noted, as some would say, under the compulsion of brute empirical fact. . . . It admits no explanation." Emergent phenomena do not have a single cause (compare *butterfly effect*) but are a product of the system as a whole. See also *uncomputable*.

ENSEMBLE FORECAST: This is a technique that makes many separate forecasts using slightly perturbed initial conditions, different models, or a combination thereof. The results are then interpreted using statistical techniques to make a probabilistic forecast. In weather forecasting, ensemble-forecast schemes were originally designed to account for the butterfly effect.

EXPONENTIAL GROWTH: A quantity grows exponentially if the rate of growth is proportional to the size of the quantity. It is often an indicator of positive feedback. Money in a bank account grows exponentially if the interest is reinvested. Note that exponential growth need not mean rapid growth (a deposit held in a chequing account, for example, grows slowly).

FEEDBACK: This term from the branch of mathematics known as control theory is used also in a variety of fields, including engineering, biology, and economics. It refers to a situation in which a portion of a signal is fed back into the source, thus modifying the signal. Feedback can be positive or negative. An example of positive feedback is a microphone pointed at a speaker—any noise is picked up by the microphone, amplified, and sent to the speaker. The speaker's output is picked up by the microphone, sent back to the speaker, and so on (until the speaker explodes or someone grabs the microphone). Positive feedback can also lead to collapse instead of growth. When a business loses customers, it has less money to invest in its products, which means it loses more customers, and so on. An example of negative feedback is the flyball governor of Chapter 3, which kept the speed of steam engines at a steady level. In general, positive feedback tends to accentuate a change to a signal, while negative feedback reduces change.

FRACTAL: This term (from the Latin *fractus*, for "broken") was coined in 1975 by Benoit Mandelbrot. Unlike the figures of classical geometry, such as circles and squares, a fractal is an irregular figure that shows a self-similar structure over a range of scales, so the same visual motifs appear no matter how close you zoom in. The figures can be generated using extremely simple mathematical procedures, but often they have a remarkably complex appearance. Fractals are better viewed than described; many examples, such as the Mandelbrot set, can be found on the Internet.

GAIA THEORY: This is the holistic theory that the earth can usefully be viewed as having the self-regulatory, homeostatic properties of a living organism.

GDP (GROSS DOMESTIC PRODUCT): The GDP is an estimate of the monetary value of goods and services produced within a nation. It does not include the so-called invisible economy, such as unpaid labour in the home, voluntary work, or the black market. Nor does it account for the costs of environmental damage to natural systems. It can therefore be highly misleading as a measure of a society's true condition.

GENE: This word (from the Greek *genos*, for "origin") refers to a rather fuzzy concept. In functional terms, it is used to describe a basic unit of inheritance that passes certain traits from generation to generation. In physical terms, it is a portion or portions of DNA that codes for a particular RNA or protein. A gene can produce different proteins if the intermediate RNA is modified.

GENOTYPE: This is the sum of the genetic information in an organism's DNA.

HOLISM: The opposite of reductionism, holism (from the Greek *holos*, meaning "whole") is the belief that certain systems cannot be understood in their constituent parts, but instead must be treated as a unified, organic whole. A holistic philosophy was espoused by artists such as Goethe in reaction to mechanistic science (and got a bad rap when the Nazi party in Germany promoted the holistic ideal of the state as a living organism).

HOMEOSTASIS: Homeostasis (from the Greek words *homeo* and *stasis*, meaning "to remain the same") is an expression coined by the American physiologist Walter Cannon in 1932. It refers to the ability of living organisms to maintain certain critical aspects of their internal state within a narrow range (as opposed to a precisely fixed level). Examples in the human body include temperature and salinity.

INITIAL CONDITION: This is the initial value of the variables at time zero for a mathematical model. In the example of the falling stone in Chapter 2, the initial condition is the starting height and velocity of the stone.

LINEAR EQUATION: An equation of the form $y = ax + b$, where a and b are constant parameters, is linear in the variable x. Linear equations are so named because when y is plotted versus x, it gives a straight line. In a linear dynamical system, the rates of change of variables are given by linear equations.

MATHEMATICAL MODEL: A mathematical model is a simulation of a physical, biological, economic, or other system using a set of equations, which typically make up a dynamical system. Mathematical models are useful for the scientific study of natural systems. Note that the dynamical system should not be confused with the underlying system, which is generally far more complicated and cannot be expressed using equations. Models can therefore provide valuable insights, but they must be used with care, especially when making predictions.

MECHANISTIC SCIENCE: This is the view that all processes, including the behaviour of living things, are machine-like and amenable to a reductionist approach. According to mechanistic science, it is possible to accurately simulate a system by building a mathematical model based on fundamental laws.

METABOLISM: This is the sum of all biochemical processes in a living organism.

MODEL ERROR: The error in a forecast that arises because of a discrepancy between the mathematical model and the underlying system is called the model error. It can be evaluated using techniques such as model drift and shadow orbits.

NAÏVE FORECAST: This is a forecast based on simple rules of thumb rather than a mathematical model (like the climatological forecast or the persistence forecast). A forecast is said to have skill if it can beat naïve forecasts.

NON-LINEAR EQUATION: Any equation that is not linear (i.e., most equations) is called non-linear. The difficulties in solving non-linear dynamical systems meant that they were little studied until the development of fast computers in the 1960s.

NORMAL DISTRIBUTION: This is a particular probability distribution; it's also known as the bell curve, for its shape. The distribution is symmetric, with 68 percent within one standard deviation of the mean and 95 percent within two standard deviations. It has many useful mathematical properties, including the central limit theorem, which states that the sum of independent, identically distributed (i.i.d.) random variables will approach a normal distribution. This means, for example, that if an error in a calculation is the result of the cumulative sum of many small, independent errors of a similar type, then it will tend to follow a normal distribution. The distribution was used by Laplace in 1783 to study measurement errors, but it has also been adopted in areas such as finance, where its use is less justified.

ODE (ORDINARY DIFFERENTIAL EQUATION): This class of equations involves mathematical derivatives, which specify the rate of change of one variable with respect to another. In the example of the falling stone in Chapter 2, the two ODEs involve the rates of change of position and velocity with time.

PANDEMIC: This is an epidemic that spreads so that it infects people over a large geographical area. In modern times, this effectively means the whole world.

PARAMETER: A parameter is a number in an equation that doesn't change with time. In the example of the falling stone in Chapter 2, the force of gravity was treated as a fixed parameter, even though in reality it varies with distance from the earth. See also *variable*.

PARAMETERIZATION: In mathematical models, parameterization refers to the representation of a complex process by an approximate equation. An example is the equations for the formation and dissipation of clouds in GCMs. A property of many models discussed in this book, including GCMs, is their sensitivity to parameterization. As a result, the models can be tuned to fit past data, but they still fail to predict the future.

PERSISTENCE FORECAST: This is a naïve forecast for a system's future based on its current state. An example of a persistence forecast would be predicting that the temperature next week will be equal to the temperature today.

PHENOTYPE: The phenotype refers to an organism's traits, such as size or skin colour; the phenotype is influenced by the genotype and the environment, whose effects work in concert.

POWER LAW: Two quantities (x and y) are said to be related by a power law if $y = cx^k$ where c and the exponent k are constant numbers. For example, if y represents the area of a circle of radius x, then x and y are related by a power law $y = \pi x^2$ with exponent 2. A power law probability distribution is one in which the probability y of a state x obeys a power law. If x represents wealth and y represents the number of people with that amount of wealth, then in many countries y approximately follows a power law relationship with exponent around -2. The negative sign means that the chances of having low wealth are much greater than the chances of having great wealth.

PROBABILISTIC FORECAST: This is a forecast expressed in terms of a probability—saying, for example, that there is an 80 percent chance it will rain tomorrow. It can be obtained by performing an ensemble forecast. Alternatively, a forecaster can modify a single model forecast into a statement of likelihood—turning a model prediction for rain into "a strong chance of rain," for example—either by mathematically accounting for expected errors in some plausible fashion or more simply by using his or her own judgment. (Of course, this is subjective, but so are many of the choices that lie behind the development of the model and any ensemble scheme.)

PROBABILITY DISTRIBUTION: This is any equation or table that assigns a probability y for any number x. For example, IQ scores were designed so that the chances of a randomly chosen person having a certain IQ x follows a normal distribution with mean 100 and standard deviation of about 15.

PROTEIN: A protein is a molecule consisting of a string of amino acids folded into a particular shape. Proteins make up the structure of cells and carry out most of an organism's functions.

RANDOM WALK: If a drunk stumbles around, taking steps of fixed length in random directions, then the path he traces out is a random walk. The random walk theory is used in physics to describe the random motion of small particles, in finance to describe the random fluctuations of an asset's value, and by the police to test drunk drivers. See also *stochastic process*.

RATIONAL NUMBERS: These are numbers that can be expressed as a fraction, such as ⅜. All other numbers (i.e., most of them) are called irrational.

REDUCTIONISM: This is the belief, championed by Descartes, that the behaviour of systems can be understood by reducing them to their simplest parts. An example is the selfish gene theory. See also *holism*.

RMS (ROOT MEAN SQUARE): For a set of N numbers $\{x_1, x_2, \Lambda, x_N\}$, the RMS is given by the square root of the sum of squares: $\sqrt{x_1^2 + x_2^2 + \Lambda + x_N^2}$. The RMS has many applications. For the case $N = 2$, it can be viewed in terms of a diagonal distance, from the theorem of Pythagoras. If the different numbers represent errors in N separate variables, the RMS is often used as a measure of the total error.

RNA (RIBONUCLEIC ACID): Created by transcription of DNA, RNA plays a key role in the cell, especially as an intermediary between genes and proteins.

SCEPTIC: One who questions and critically examines whatever passes for knowledge, and acknowledges doubt and uncertainty. A global-warming sceptic should therefore be someone who believes that we cannot predict climate change, rather than (its current usage) someone who believes it will not happen.

SELFISH GENE THEORY: This is the reductionist theory, first proposed by Richard Dawkin (1976), that evolution is the net product of independent, individual genes trying to maximize their reproductive success. The theory is therefore a kind of efficient market theory for the genome (see EMH). Supporters of the theory emphasize that it does not imply that genes are endowed with their own motives—only that they can be viewed as acting as if they are.

SENSITIVITY TO INITIAL CONDITION: A mathematical model has this property if small changes to the initial value of the variables (i.e., the initial condition) result in very different forecasts. See also *butterfly effect*.

SENSITIVITY TO PARAMETERIZATION: A mathematical model has this property if small changes to the parameterizations used (for example, the value of certain parameters) result in very different forecasts. Climate change predictions are generally insensitive to small changes in initial condition but sensitive to parameterization.

SHADOW ORBIT: This refers to a trajectory of a dynamical system that stays within a specified distance—the shadow radius—of a sequence of observations. A model's accuracy over a certain time period can be assessed by searching for a trajectory, starting from a perturbed initial condition, which shadows within as small a shadow radius as possible (while allowing for uncertainty in the observations). The smaller the shadow radius, the better the model is at matching the data. The expected radius can be compared with estimates from the model drift. Together, they provide two independent methods to assess model error.

STANDARD DEVIATION: This is a measure of the width of a normal distribution. It's also used as a measure of volatility of some fluctuating quantity, where the implicit assumption is that the deviations from the mean follow a normal distribution. The square of the standard deviation is known as the variance.

STOCHASTIC: A process is stochastic if it is random. If two players bet repeatedly on the flip of a coin, then the score will represent a stochastic process, with some similarity to the fluctuations of a financial asset.

TRAJECTORY: A trajectory is the solution to a dynamical system, starting from a particular initial condition, which traces out the values of the variables as a function of time. It's sometimes also called an orbit (as in shadow orbit).

UNCOMPUTABLE: This term is used here to refer to a physical, biological, or other real-world system that cannot be accurately modelled using equations. Some features of uncomputable systems may be compared with the emergent properties of complex mathematical systems. An example is the formation or dissipation of clouds, for which no accurate equations exist.

VALIDATION: An argument is considered valid if it is free from any obvious logical flaws. By analogy, a mathematical model is valid if there isn't anything obviously wrong with it. The word is confusingly used in areas such as climate science; people say, for example, that a model has been "validated" because its results are consistent with a set of observations. This "validity" is then used to imply that the model is an accurate representation of the underlying system. However, the model equations can always be tuned to fit observed data, and no matter what tests are applied, there is no guarantee that a model will continue to work in the future, or that it even bears any close resemblance to the function of the system (Oreskes et al. 1994). Note that even highly imperfect models can still be useful for gaining insights and proposing testable hypotheses.

VARIABLE: A variable is any number in an equation that is allowed to change. In the example of the falling stone in Chapter 2, time, position, and velocity are all variables. See also *parameter*.

VOLATILITY: In economics, volatility refers to the size of fluctuations in an asset's value around its mean. It's often used as a proxy for risk. Volatility can be measured by the standard deviation, though this assumes that the deviations follow a normal distribution.

Bibliography

Abraham, C. (2005). Race. *The Globe and Mail,* June 18, 2005.

Abraham, R. (1994). *Chaos, Gaia, Eros: A chaos pioneer uncovers the three great streams of history.* New York: HarperCollins.

Acar, M., Becskei, A. and van Oudenaarden, A. (2005). Enhancement of cellular memory by reducing stochastic transitions. *Nature, 435,* 228–232.

ACIA (2004). *Impacts of a warming arctic: Arctic climate impact assessment.* Cambridge: Cambridge University Press.

Ackerman, J. (2001). *Chance in the house of fate: A natural history of heredity.* Boston: Houghton Mifflin.

Albright, R. (2002). What can past technology forecasts tell us about the future? *Technological Forecasting and Social Change, 69,* 443–64.

Allen, M. (2005). A novel view of global warming. *Nature, 433,* 198.

Allen, M., Kettleborough, J., and Stainforth, D. (2002). Model error in weather and climate forecasting. From the proceedings of the 2002 ECMWF Predictability Seminar, European Centre for Medium-range Weather Forecasting, Reading, U.K., pp. 275–94.

Alter, D., and Eny, K. (2005). The relationship between the supply of fast-food chains and cardiovascular outcomes. *Canadian Journal of Public Health, 96* (3), 173–77.

Anonymous (1999). We woz wrong. *The Economist,* December 16, 1999.

Anonymous (1999a). Rethinking thinking. *The Economist,* December 16, 1999.

Anonymous (1999b). It never rains. *The Economist,* September 16, 1999.

Anonymous (2000). Who owns your genes? *The Economist,* June 29, 2000.

Anonymous (2000a). Climate change may be down to farming. *Guardian,* November 4, 2000.

Anonymous (2001). The "skeptical environmentalist." *The Economist,* September 6, 2001.

Anonymous (2001a). Say "R". *The Economist,* November 29, 2001.

Anonymous (2002). The race to computerise biology. *The Economist,* December 12, 2002.

Anonymous (2003). Hot potato revisited. *The Economist,* November 6, 2003.

Anonymous (2004). Up close, and personal. *The Economist,* October 16, 2004.

Anonymous (2005). Are you being served? *The Economist,* April 21, 2005.

Anonymous (2005a). Amazon destruction accelerating in Brazil. *Reuters,* May 19, 2005.

Anonymous (2005b). Divining the future. *The Economist,* January 15, 2005.

Anonymous (2005c). Q&A with Laurie Garrett. *Foreign Affairs, 84* (July/August).

Anonymous (2005d). Hotting up. *The Economist,* February 3, 2005.

Anonymous (2005e). WHO: Impossible to predict bird flu deaths. *The Associated Press,* September 30, 2005. [http://abcnews.go.com/Health/wireStory?id=1172896]

Anonymous (2005f). Weather risk: Natural hedge. *The Economist,* September 29, 2005.

Anonymous (2005g). Personalised drugs 'decades away.' BBC News Online, September 21, 2005. <http://news.bbc.co.uk/1/hi/health/4267304.stm>.

Aristotle (1952). *Meteorologica.* Translation by Henry D. P. Lee. London: Heinemann.

Aristotle (1981). *The Politics.* Translated by T. J. Saunders and revised by T. A. Sinclair. London: Penguin Classics.

Arrhenius, S. (1896). On the influence of carbonic acid in the air upon the temperature of the ground. *Philosophical Magazine, 41,* 237.

Asimov, I. (1951). *Foundation.* New York: Avon.

Bachelier, L. (1900). Théorie de la spéculation. *Annales Scientifiques de l'Ecole Normale Supérieure,* 17, 21–86. [English translation by P. Cootner (1964), pp. 17–78.]

Baker, D., DeLong, J. B., and Krugman, P. R. (2005). Asset returns and economic growth. *Brookings Papers on Economic Activity, 1,* 289–315.

Barabási, A. L., and Albert, R. (1999). Emergence of scaling in random networks. *Science, 286,* 509–12.

Barabási, A. L., and Oltvai, Z. N. (2004). Network biology: Understanding the cell's functional organization. *Nature Reviews Genetics, 5,* 101–13.

Barnett, J., and Adger, W. M. (2003). Climate dangers and atoll countries. *Climatic Change, 61,* 321.

Baron-Cohen, S. (2003). *The essential difference: Male and female brains and the truth about autism.* New York: Basic Books.

Bass, T. A. (1999). *The predictors.* New York: Henry Holt.

Batra, R. (1987). *The great depression of 1990.* New York: Simon and Schuster.

Bazell, R. (2005). Movies help doctors discover autistic minds: Cutting-edge research at Yale may help with early detection. *NBC Nightly News,* February 24, 2005.

Bernstein, P. L. (1998). *Against the gods: The remarkable story of risk.* Toronto: John Wiley.

Bjerknes, V. (1904). Das Problem der Wettervorhersage, betrachtet vom Stadpunkte der Mechanik und der Physik. (Weather forecasting as a problem in mechanics and physics). *Meteorologische Zeitschrift, 21,* 1–7.

Bjerknes, V. (1911). *Dynamic meteorology and hydrography, Part 2: Kinematics.* New York: Gibson Bros.

Blum, W. (2002). Folgenloser Flügelschlag. *Bild der wissenschaft,* June 2005, p. 46.

Bogen, J. E. (1975). Some educational aspects of hemisphere specialization. *UCLA Educator, 17,* 24–32.

Bohm, D. (1974). On the subjectivity and objectivity of knowledge. In John Lewis (ed.), *Beyond chance and necessity.* London: Garnerstone Press.

Bollerslev, T. P. (1986). Generalized autoregressive conditional heteroskedasticity. *Journal of Econometrics, 31,* 307–27.

Bosart, L. F. (2003). Whither the weather analysis and forecasting process? *Weather Forecasting, 18,* 520–29.

Bragg, M. (1999). *On giants' shoulders.* New York: Wiley.

Brookes, M. (2004). *Extreme measures: The dark visions and bright ideas of Francis Galton.* New York: Bloomsbury.

Brooks, R. (2002). *Flesh and machines: How robots will change us.* New York: Pantheon.

Brumfiel, G. (2004). Newton's religious screeds get online airing. *Nature, 430,* 819.

Buchanan, M. (2000). *Ubiquity.* London: Weidenfeld and Nicolson.

Buizza, R., Barkmeijer, J., Palmer, T. N., and Richardson, D. S. (2000). Current status and future developments of the ECMWF Ensemble Prediction System. *Meteorological Applications, 7,* 163–75.

Byers, M. (2005). On thinning ice. *London Review of Books, 27,* January 6, 2005.

Byrne, F. (2005). Lunch with the FT: Make it snappy. *Financial Times*, January 28, 2005.

Calder, N. (2003). *Magic universe: The Oxford guide to modern science*. Oxford: Oxford University Press.

Campbell, D. and Lee, R. (2003). "Carnage by computer": The blackboard economics of the 2001 foot and mouth epidemic. *Social & Legal Studies*, 12, 425-459.

Capra, F. (1983). *The tao of physics*. Boston: Shambhala.

Capra, F. (1996). *The web of life: A new scientific understanding of living systems*. New York: Anchor Books.

Capra, F. (2002). *The hidden connections*. New York: Anchor Books.

Chagnon, S., ed. (2000). *El Niño 1997–1998: The climate event of the century*. Oxford: Oxford University Press.

Charlson, R., Lovelock, J., Andreae, M., and Warren, S. (1987). Oceanic phytoplankton, atmospheric sulphur, cloud albedo and climate. *Nature*, *326*, 655–61.

Cheung, Y.-W., Chinn, M. and Garcia Pascual, A. (2005). Empirical exchange rate models of the nineties: are any fit to survive? *Journal of International Money & Finance, 24*, 1150–1175.

Chomsky, N. (1988). *Language and the problems of knowledge*. Cambridge, MA: MIT Press.

Christianson, J.R. (2000). *On Tycho's island: Tycho Brahe and his assistants 1570–1601*. Cambridge: Cambridge University Press.

Churchill, W. (1932). Fifty years hence. *Popular Mechanics* (March), 390.

Cohn, S. (1997). An introduction to estimation theory. *Journal of the Meteorological Society of Japan, 75*, 257–88.

Connor, J. A. (2004). *Kepler's witch : An astronomer's discovery of cosmic order amid religious war, political intrigue, and the heresy trial of his mother*. New York: HarperCollins.

Cootner, P. (1964). *The random character of stock market prices*. Cambridge, MA: MIT Press.

Cornford, F. M. 1969 (1923). *Greek religious thought*. New York: AMS Press.

Coulson, S., and Williams, R. F. (2005). Hemispheric asymmetries and joke comprehension. *Neuropsychologia, 43*, 128–41.

Coupland, D. (2002). *Souvenir of Canada*. Vancouver: Douglas and McIntyre.

Cowles, A. (1933). Can stock market forecasters forecast? *Econometrica, 12*, 206–14.

Cox, J. D. (2002). *Storm watchers: The turbulent history of weather prediction from Franklin's kite to El Niño*. Hoboken, NJ: John Wiley.

Crichton, M. (2002). *Prey.* New York: HarperCollins.

Crichton, M. (2004). *State of Fear.* New York: HarperCollins.

Cullen, M.R. (2005). Serum osteopontin levels—is it time to screen asbestos-exposed workers for pleural mesothelioma? *New England Journal of Medicine, 353,* 1564–73.

Daly, H.E. and Cobb, J.B. Jr. (1989). *For the common good.* Boston: Beacon Press.

Danchin, A. (2002). *The delphic boat: What genomes tell us.* Cambridge, MA: Harvard University Press.

Danforth, C., and Yorke, J. A. (2005). Making forecasts for chaotic physical processes. *Physical Review Letters, 96,* 144102.

Darwin, C. (1905). *The voyage of a naturalist round the world in H.M.S. "Beagle".* New York: Routledge.

Dawkins, R. (1976). *The selfish gene.* Oxford: Oxford University Press.

De Atauri, P., Orrell, D., Ramsey, S., and Bolouri, H. (2004). Evolution of "design principles" in biology and engineering. *IEE Systems Biology, 1,* 28–40.

Deacon, R. (1968). *John Dee: Scientist, geographer, astrologer and agent to Elizabeth I.* London: Frederick Muller.

Dennis, C. (2004). The most important sexual organ. *Nature, 427,* 390.

Descartes, R. (1960). *Discourse on Method.* Translation by Arthur Wollaston. Harmondsworth, UK: Penguin.

Diamond, J. (2005). *Collapse.* New York: Penguin.

Downton, R. A., and Bell, R. S. (1988). The impact of analysis differences on a medium-range forecast. *Meteorological Magazine, 117,* 279–85.

Draper, J. W. (1874). *History of the conflict between religion and science.* New York: D. Appleton and Company.

Drexler, E. (1986). *Engines of creation.* New York: Anchor Books.

Duncan, D. E. (2003). Reversing bad truths. *Discover, 24,* 20.

Dunphy, S. (2004). The inflation conundrum. *Seattle Times,* February 22, 2004.

Ebert, E. E., Damrath, U., Wergen, W., and Baldwin, M. E. (2003). The WGNE assessment of short-term quantitative precipitation forecasts. *Bulletin of the American Meteorological Society, 84,* 481–92.

Edelman, G. (1987). *Neural Darwinism: The theory of neuronal group selection.* New York: Basic Books.

Edwards, B. (1999). *Drawing on the right side of the brain.* London: HarperCollins.

Edwards, P. N. (2000). A brief history of atmospheric general circulation modeling. In David A. Randall (ed.), *General circulation model development*. San Diego: Academic Press.

Ehrlich, P. R. (1968). *The population bomb*. New York: Ballantine Books.

Ehrlich, P. R. (2000). *Human natures: Genes, cultures, and the human prospect*. Washington, DC: Island Press.

Ehrlich, P. R., and Ehrlich, A. (2004). *One with Nineveh: Politics, consumption and the human future*. Washington, DC: Island Press.

Emanuel, K. (2005). Increasing destructiveness of tropical cyclones over the past 30 years. *Nature, 436*, 686–88.

Engle, R. F. (1982). Autoregressive conditional heteroskedasticity with estimates of the variance of UK inflation. *Econometrica, 50*, 987–1008.

Errico, R. M., Langland, R., and Baumhefner, D. P. (2002). The workshop in atmospheric predictability. *Bulletin of the American Meteorological Society, 83*, 1341–43.

Eto, M., Watanabe, K., and Makino, I. (1989). Increased frequencies of apolipoprotein E2 and E4 alleles in patients with ischemic heart disease. *Clinical Genetics, 36*, 183–88.

Eubank, S., et al. (2004). Modelling disease outbreaks in realistic urban social networks. *Nature, 429*, 180–84.

Fama, E. F. (1965). Random walks in stock-market prices. *Selected Papers, 16*. Chicago: University of Chicago, Graduate School of Business.

Ferguson, G. W. (1954). *Signs and symbols in Christian art*. New York: Oxford University Press.

Field, G. C. (1956). *The philosophy of Plato*. Oxford: Oxford University Press.

Fischetti, M. (2001). Drowning New Orleans. *Scientific American*, October 2001.

Foley, S. (2005). Roche is boosted by avian flu fears. *Independent*, October 20, 2005.

Frank, K. T., Petrie, B., Choi, J. S., and Leggett, W. C. (2005). Trophic cascades in a formerly cod-dominated ecosystem. *Science, 10*, 1621–23.

Freese, B. (2003). *Coal: A human history*. Cambridge, MA: Persius.

Frisinger, H. H. (1977). *The history of meteorology to 1800*. New York: Science History Publications.

Galilei, G. (1967). *Dialogue concerning the two chief world systems: Ptolemaic and Copernican*. Translated by Stillman Drake. Berkeley: University of California Press.

Galton, F. (1865). Hereditary talent and character. *MacMillan's Magazine,* *12,* 157–66, 318–27.

Galton, F. (1869). *Hereditary genius: An inquiry into its laws and consequences.* London: Macmillan.

Galton, F. (1871). Experiments in pangenesis, by breeding from rabbits of a pure variety, into whose circulation blood taken from other varieties had previously been largely transfused. *Proceedings of the Royal Society, 19,* 393–410.

Galton, F. (1876). On the height and weight of boys aged 14, in town and country public schools. *Journal of the Anthropological Institute, 6,* 174–81.

Galton, F. (1878). Composite portraits. *Journal of the Anthropological Institute, 8,* 132–44.

Galton, F. (1886). Regression toward mediocrity in hereditary stature. *Journal of the Anthropological Institute, 15,* 246–63.

Galton, F. (1888). Co-relations and their measurements, chiefly from anthropometric data. *Proceedings of the Royal Society, 45,* 135–45.

Galton, F. (1891). Identification by fingerprints. *Nineteenth Century, 30,* 303–11.

Galton, F. (1898). A diagram of heredity. *Nature, 57,* 293.

Galton, F. (2001). *The art of travel; or, shifts and contrivances available in wild countries.* London: Phoenix Press.

Garrett, L. (1994). *The coming plague.* New York: Farrar, Straus, and Giroux.

Garrett, L. (2005). The next pandemic? *Foreign Affairs, 84* (July/August).

Gibbs, W. W. (2001). Cybernetic cells. *Scientific American, 285* (August), 54–57.

Giles, L. C., Glonek, G. F. V., Luszcz, M. A., and Andrews, G. R. (2005). The effect of social networks on 10-year survival in very old Australians: The Australian longitudinal study of ageing. *Journal of Epidemiology and Community Health, 59,* 574–79

Gill, P. E., Murray, W., and Wright, M. H. (1981). *Practical optimization.* London: Academic Press.

Gillham, N. W. (2001). *A life of Sir Francis Galton: From African exploration to the birth of eugenics.* Oxford: Oxford University Press.

Gilmour, I. (1998). Nonlinear model evaluation: ι-shadowing, probabilistic prediction and weather forecasting. Ph.D. thesis, Oxford University.

Gladwell, M. (2004). The picture problem: Mammography, air power, and the limits of looking. *The New Yorker,* December 13, 2004.

Gladwell, M. (2005). *Blink: The power of thinking without thinking.* New York: Little, Brown.

Glaser, R., Rice, J., Sheridan, J., Fertel, R., Stout, J., Speicher, C., Pinsky, D., Kotur, M., Post, A., Beck, M., and Kiecolt-Glaser, J. (1987). Stress-related immune suppression: Health implications. *Brain, Behavior, and Immunity, 1*, 7–20.

Gleick, J. *Chaos: Making a new science.* (1987). London: Sphere Books.

Goldstein, D.B. and Cavalleri, G.L. (2005). *Genomics: Understanding human diversity. Nature, 437*, 1241–1242.

Gonzalez, E., Kulkarni, H., Bolivar, H., et al. (2005). The influence of CCL3L1 gene-containing segmental duplications on HIV-1/AIDS susceptibility. *Science, 307*, 1434-40.

Gould, S. J. (1977). *Ontogeny and phylogeny.* Cambridge, MA: Belknap Press.

Greenblatt, S. (2004). *Will in the world: How Shakespeare became Shakespeare.* New York: W. W. Norton.

Greene, L. (2005). Pill may contain, not cure bird flu. *St. Petersburg Times,* October 31, 2005.

Gribbin, J. (2002). *The scientists: A history of science told through the lives of its greatest inventors.* New York: Random House.

Griffiths, J. (2004). *A sideways look at time.* New York: Tarcher/Penguin.

Gunson, P. (1999). 20,000 feared dead in Caribbean resorts buried under tide of mud. *Independent,* December 21, 1999.

Guthrie, W. K. C. (1962–1981). *A history of Greek philosophy.* 6 vols. Cambridge: Cambridge University Press.

Gutkin, S. (1999). Pointing fingers in Venezuela. *The Associated Press,* December 29, 1999. [http://www.cbsnews.com/stories/1999/12/20/world/main141926.shtml]

Haga, S. B., Khoury, M. J. and Burke, W. (2003). Genomic profiling to promote a healthy lifestyle: not ready for prime time. *Nature Genetics, 34*, 347–350.

Halpern, P. (2000). *The pursuit of destiny: A history of prediction.* Cambridge, MA: Perseus.

Hamill, T. M., Hansen, J. A., Mullen, S. L., and Snyder, C. (2006). Meeting summary: Workshop on ensemble forecasting in the short to medium range. *Bulletin of the American Meteorological Society, 87* (in press).

Hammock, E., and Young, L. J. (2005). Microsatellite instability generates diversity in brain and sociobehavioral traits. *Science, 308*, 1630–34.

Hansen, J., Nazarenko, L., et al. (2005). Earth's energy imbalance: Confirmation and implications. *Science, 308*, 1431–35.

Harrison, M. S. J., Palmer, T. N., Richardson, D. S., and Buizza, R. (1999). Analysis and model dependencies in medium-range ensembles: Two transplant case studies. *Quarterly Journal of the Royal Meteorological Society, 126,* 2487–515.

Harvell, C. D., et al. (2002). Climate warming and disease risks for terrestrial and marine biota. *Science, 296,* 2158–62.

Harvey, F., and Cookson, C. (2005). US defends stance on climate change. *Financial Times,* June 16, 2005.

Haugen, R. A., and Lakonishok, J. (1992). *The incredible January effect.* New York: Irwin.

Hawken, P. (1994). *The ecology of commerce.* New York: HarperCollins.

Heath, J. R., Phelps, M. E., and Hood, L. (2003). Nanosystems biology. *Molecular Imaging and Biology, 5,* 312–25.

Herrigel, E. (1953). *Zen in the art of archery.* New York: Random House.

Highfield, R. (2004). Tsunamis can be predicted if governments are motivated. *The Daily Telegraph,* December 29, 2004.

Hillman, J. (1975). *Re-Visioning psychology.* New York: Harper Perennial.

Hinkle, G. J. (1996). Marine salinity: Gaian phenomenon? In P. Bunyard (ed.), *Gaia in Action.* Edinburgh: Floris Books.

Hirsch, M. W. (1984). The dynamical systems approach to differential equations. *Bulletin of the American Mathematical Society, 11,* 1–64.

Hoegh-Guldberg, O. (1999). Coral bleaching, climate change and the future of the world's coral reefs. *Marine Freshwater Research, 50,* 839–66.

Holton, J. R. (1992). *An introduction to dynamic meteorology.* London: Academic Press.

Homer-Dixon, T. F. (2000). *The ingenuity gap.* Toronto: Vintage.

Horgan, J. (1996). *The end of science.* New York: Addison-Wesley.

Horgan, J. (2004). Do our genes influence behavior?: Why we want to think they do. *Chronicle of Higher Education,* November 26, 2004.

Houtekamer, P. L., Mitchell, H. L., Pellerin, G., Buehner, M., Charron, M., Spacek, L., and Hansen, B. (2005). Atmospheric data assimilation with an ensemble Kalman filter: Results with real observations. *Monthly Weather Review, 133,* 604–20.

Hubbard, R., and Wald, E. (1999). *Exploding the gene myth.* Boston: Beacon Press.

Huber, P. W., and Mills, M. P. (2005). *The bottomless well.* New York: Basic Books.

Iamblichus, (1918). *Life of Pythagoras.* Translated by Thomas Taylor. Hollywood, CA: Krotona.

Ingram, J. (2005). *Theatre of the mind.* Toronto: HarperCollins.

Intergovernmental Panel on Climate Change. (2000). *Emissions scenarios: A special report of working group III of the Intergovernmental Panel on Climate Change.* Edited by N. Nakicenovic and R. Swart. Cambridge: Cambridge University Press.

Intergovernmental Panel on Climate Change. (2001). *Climate change 2001: Impacts, adaptation and vulnerability.* Edited by J. J. McCarthy et al. Cambridge: Cambridge University Press.

Intergovernmental Panel on Climate Change. (2001a). *Climate change 2001: The scientific basis.* Edited by J. T. Houghton, Y. Ding, D. J. Griggs, M. Noguer, P. J. van der Linden, and D. Xiaosu. Cambridge: Cambridge University Press.

International Human Genome Sequencing Consortium (2001). Initial sequencing and analysis of the human genome. *Nature, 409,* 860–921.

Jack, A., Lau, J., Tucker, S., and Hille, K. (2005). Human flu preparation would cost billions. *Financial Times,* October 21, 2005.

Jackson, R. J., Ramsay, A. J., Christensen, C. D., Beaton, S., Hall, D. F., and Ramshaw, I. A. (2001). Expression of mouse interleukin-4 by a recombinant ectromelia virus suppresses cytolytic lymphocyte responses and overcomes genetic resistance to mousepox. *Journal of Virology, 75,* 1205–10.

Jaenicke, R. (2005). Abundance of cellular material and proteins in the atmosphere. *Science, 308, 73.*

Jasanoff, S., and Wynne, B. (1998). Science and decision-making. In S. Rayner and E. L. Malone (eds.), *Human choice and climate change. Volume 1: The societal framework.* Columbus, OH: Battelle Press.

Jardine, L. (1999). *Ingenious pursuits: Building the scientific revolution.* New York: Nan A. Talese/Doubleday.

Jevons, W. S. (1888). *The Theory of Political Economy.* London: Macmillan and Co.

Jewson, S., Ziehman, C., and Brix, A. (2002). Use of meteorological forecasts in weather derivative pricing. In B. Dickle (ed.), *Climate risk and the weather market: Financial risk management with weather hedges.* London: Risk Books.

Johnson, R. (2005). A genius explains. *The Guardian,* February 12, 2005.

Jordan, N. (1993). *A Neil Jordan reader.* New York: Vintage Books.

Junger, S. (1997). *The perfect storm.* London: Fourth Estate.

Kahn, H., and Wiener, A. J. (1967). *The year 2000: A framework for speculation on the next thirty-three years.* New York: Macmillan.

Kamboh, M. I. (1995). Apolipoprotein E polymorphism and susceptibility to Alzheimer's disease. *Human Biology, 67,* 195–215.

Katz, R. W. (2002). Sir Gilbert Walker and a connection between El Niño and statistics. *Statistical Science, 17,* 97–112.

Keepin, B., and Wynne, B. (1984). Technical analysis of IIASA energy scenarios. *Nature, 312,* 691–95.

Kehoe, T. J. (2005). An Evaluation of the Performance of Applied General Equilibrium Models of the Impact of NAFTA. In T. J. Kehoe, T. N. Srinivasan, and John Whalley (eds.), *Frontiers in Applied General Equilibrium Modeling: Essays in Honor of Herbert Scarf.* Cambridge: Cambridge University Press, 341–77.

Keller, E. F. (1985). *Reflections on gender and science.* New Haven, CN: Yale University Press.

Keller, E. F. (2000). *The century of the gene.* Cambridge, MA: Harvard University Press.

Keller, E. F. (2002). *Making sense of life.* Cambridge, MA: Harvard University Press.

Kennedy, D. (2004). Climate change and climate science. *Science, 304,* 1565.

Kerr, R. A. (2003). Another way to take the ocean's pulse. *Science, 299,* 337.

Kerr, R. A. (2004). Second thoughts on skill of El Niño predictions. *Science, 290,* 257–58.

Kerr, R. A. (2004a). Storm-in-a-box forecasting. *Science, 304,* 946–48.

Kerr, R. A. (2004b). Three degrees of consensus. *Science, 305,* 932–34.

Kerr, R. A. (2005). How hot will the greenhouse world be? *Science, 309,* 100.

Kevles, D. J., and Hood, L., eds, (1992). *The code of codes: Scientific and social issues in the human genome project.* Cambridge, MA: Harvard University Press.

Keynes, J. M. (1937). The General Theory of Employment. *Quarterly Journal of Economics, 51,* 209–23.

Kirby, A. (2004). World "appeasing" climate threat. BBC News Online, June 3, 2004. <http://news.bbc.co.uk/2/hi/science/nature/3766831.stm>.

Kirch, P. V. (1984). *The evolution of the Polynesian chiefdoms.* Cambridge: Cambridge University Press.

Klin, A., Jones, W., Schultz, R., Volkmar, F., and Cohen, D. (2002). Defining and quantifying the social phenotype in autism. *American Journal of Psychiatry, 159,* 895–908.

Knight, F. H. (1921). *Risk, uncertainty, and profit.* Boston: Houghton Mifflin.

Koestler, A. (1968). *The sleepwalkers.* London: Hutchinson.

Kolbert, E. (2005). The climate of man, part 1. *The New Yorker,* April 25, 2005.

Kollar, E. J., and Fisher, C. (1980). Tooth induction in chick epithelium: Expression of quiescent genes for enamel synthesis. *Science, 207,* 993–95.

Kosko, B. (1993). *Fuzzy thinking: The new science of fuzzy logic.* New York: Hyperion.

Koutsogeorgopoulou, V. (2000). A post-mortem on economic outlook projections. Working paper 274, Economics department, Organization for Economic Co-operation and Development.

Krumbein, W. E., and Lapo, A. V. (1996). Vernadsky's biosphere as a basis of geophysiology. In P. Bunyard (ed.), *Gaia in Action.* Edinburgh: Floris Books.

Kuhn, T. S. (1962). *The structure of scientific revolutions.* Chicago: University of Chicago Press.

Kurzweil, R., and Grossman, T. (2004). *Fantastic voyage: Live long enough to live forever.* New York: Rodale Books.

Lawlor, R. (1982). *Sacred geometry.* New York: Thames and Hudson.

Leemans, R., and Eickhout, B. (2004). Another reason for concern: Regional and global impacts on ecosystems for different levels of climate change. *Global Environmental Change (Part A), 14,* 219–28.

Lerner, G. (1987). *The creation of patriarchy.* Oxford: Oxford University Press.

Leslie, A. M., and Keeble, S. (1987). Do six-month-old infants perceive causality? *Cognition, 25,* 265–88.

Lewontin, R. C. (1991). *Biology as ideology: The doctrine of DNA.* Toronto: Anansi.

Lewontin, R. C. (2001). *It ain't necessarily so: The dream of the human genome and other illusions.* New York: New York Review of Books.

Liu, Y., et al. (1999). Statistical properties of the volatility of price fluctuations. *Physical Review E, 60,* 1–11.

Lively, P. (2005). As mad as it gets. *The Sunday Times,* April 10, 2005.

Lomborg, B. (2001). *The skeptical environmentalist: Measuring the real state of the world.* Cambridge: Cambridge University Press.

Longini, I. M., et al. (2004). Containing pandemic influenza with antiviral agents. *American Journal of Epidemiology, 159,* 623–33.

Loomes, G. C. (1998). Probabilities vs money: A test of some fundamental assumptions about rational decision making. *Economic Journal, 108,* 477–89.

Lorenz, E. N. (1963). Deterministic nonperiodic flow. *Journal of Atmospheric Science, 20,* 130–41.

Lorenz, E. N. (1982). Atmospheric predictability experiments with a large numerical model. *Tellus, 34,* 505–13.

Lorenz, E. N. (1996). Predictability: A problem partly solved. In T. Palmer (ed.), *Predictability*. Reading, U.K.: European Centre for Medium-range Weather Forecasting.

Lovelock, J. E. (2006). *The revenge of Gaia*. London: Penguin.

Lovelock, J. E. (2000). *Homage to Gaia*. Oxford: Oxford University Press.

Lovelock, J. E. (1991). *Healing Gaia*. New York: Harmony Books.

Lovelock, J. E. (1979). *Gaia: A new look at life on Earth*. Oxford: Oxford University Press.

Lux, T., and Marchesi, M. (1999). Scaling and criticality in a stochastic multi-agent model of a financial market. *Nature, 397,* 498–500.

Machamer, P., ed. (1998). *The Cambridge companion to Galileo*. Cambridge: Cambridge University Press.

Mackay, C. (1852). *Extraordinary popular delusions and the madness of crowds*. London: National Illustrated Library. (Reprinted in 1995 by Wordsworth Editions, Hertfordshire, UK.)

Maddox, B. (2002). *Rosalind Franklin: The dark lady of DNA*. New York: HarperCollins.

Malakoff, D. (1997). Thirty Kyotos needed to control warming. *Science, 278,* 2048.

Malkiel, B. (1999). *A random walk down Wall Street*. New York: W. W. Norton.

Manabe, S., et al. (1965). Simulated climatology of general circulation with a hydrologic cycle. *Monthly Weather Review, 93,* 769–98.

Mandelbrot, B. (1967). How long is the coast of Great Britain?: Statistical self-similarity and fractional dimension. *Science, 155,* 636–38.

Mandelbrot, B., and Hudson, R. L. (2004). *The (mis)behavior of markets: A fractal view of risk, ruin and reward*. New York: Basic Books.

Mann, C. (1991). Lynn Margulis: Science's unruly earth mother. *Science, 252,* 378–81.

Manning, J. T. (2002). *Digit ratio: A pointer to fertility, behavior and health*. Piscataway, NJ: Rutgers University Press.

Manning, J. T., Baron-Cohen, S., Wheelwright, S., and Sanders, G. (2001). The 2nd to 4th digit ratio and autism. *Developmental Medicine and Child Neurology, 43,* 160–64.

Margolis, J. *A brief history of tomorrow*. London: Bloomsbury.

Marmot, M. G., Davey Smith, G., Stansfield, S., Patel, C., North, F., and Head, J. (1991). Health inequalities among British civil servants: The Whitehall II study. *Lancet, 337,* 1387–93.

Martin, G. and Loeb, L. (2004). Ageing: Mice and mitochondria. *Nature, 429,* 357–359.

Matthews, R. (2000). The bizarre case of the chromosome that never was. *Sunday Telegraph,* May 14, 2000.

Matthews, R. (2001). Don't blame the butterfly. *New Scientist, 171,* 27.

Matthews, R. (2004). Science turns to philosophy in search for truth. *Daily Telegraph,* July 7, 2004.

Matthews, R., and Fraser, L. (2000). After years of inquiry no-one knows how many lives vCJD will claim. *Daily Telegraph,* October 29, 2000.

Mattick, John S. (2005). The functional genomics of noncoding RNA. *Science, 309,* 1525–26.

Maturana, H. R., and Varela, F. J. (1987). *The tree of knowledge.* Boston: Shambala.

Maunder, M. (1997). The conservation of the extinct toromiro tree. *Curtis's Botanical Magazine, 14,* 226–31.

Maxwell, J. C. (1868). On governors. *Proceedings of the Royal Society of London, 16,* 270–83.

McCarthy, G. (2004). The price of living in urbanized areas. *The San Bernardino Sun,* June 29, 2004.

McCauley, J. L. (2004). *Dynamics of markets: Econophysics and finance.* Cambridge: Cambridge University Press.

McCusker, J. (2004). Forecast calls for accuracy, less chaos. *Everett Herald,* April 11, 2004.

McKibben, B. (2003). *Enough.* New York: Times Books.

McKitrick, R. R. (1998). The Econometric Critique of Applied General Equilibrium Modelling: The Role of Functional Forms. *Economic Modelling,* 15, 543-573.

McMurdy, D. (2004). Bakers cope with rising ingredient costs. *Edmonton Journal,* November 20, 2004.

McNees, S. K., and Ries, J. (1992). How large are economic forecast errors? *New England Economic Review* (July–August).

Meadows, D. H., Meadows, D. L., Randers, J., and Behrens, W. (1972). *The limits to growth.* New York: Universe Books.

Meadows, D. H., Meadows, D. L., and Randers, J. (1992). *Beyond the limits.* Toronto: McClelland andStewart.

Meehl, G. A., and Tebaldi, C. (2004). More intense, more frequent, and longer-lasting heat waves in the 21st century. *Science, 305,* 994–97.

Midgley, M. (1985). *Evolution as a religion: Strange hopes and stranger fears.* London: Methuen.

Midgley, M. (2000). *Gaia: The next big idea.* London: Demos.

Millennium Ecosystem Assessment (2005). *Ecosystems and human well-being: Desertification synthesis.* Washington, DC: World Resources Institute.

Moore, D. S. (2002). *The dependent gene: The fallacy of "nature vs. nurture."* New York: Times Books.

Moran, J. M., Morgan, M. D., and Pauley, P. M. (1997). *Meteorology: The atmosphere and the science of weather.* Upper Saddle River, NJ: Prentice Hall.

Morgan, T. H., Sturtevant, A. H., Muller, H. J., and Bridges, C. B. (1915). *The mechanism of Mendelian heredity.* New York: Henry Holt.

Morohashi, M., Winn, A. E., Borisuk, M. T., Bolouri, H., Doyle, J., and Kitano, H. (2002). Robustness as a measure of plausibility in models of biochemical networks. *Journal of Theoretical Biology, 216,* 19–30.

Moyers, B. (2004). On receiving Harvard Medical School's Global Environment Citizen Award. Commondreams.org, December 6, 2004. <http://www.commondreams.org/views04/1206–10.htm>.

Muir, H. (2003). Einstein and Newton showed signs of autism. *New Scientist* (April).

Murphy, J. M., Sexton, D. M. H., Barnett, D. N., Jones G. S.,Webb, M. J., et al (2004). Quantifying uncertainties in climate change using a large ensemble of global climate model predictions. *Nature, 430,* 768–72.

Murphy, R. C. (1926). Oceanic and climatic phenomena along the west coast of South America during 1925. *Geographical Review, 16,* 26–54.

National Academy of Sciences, Committee on Atmospheric Sciences Panel on Weather and Climate Modification (1966). *Weather and climate modification: Problems and prospects,* vol. 2. Washington, DC: National Academy of Sciences.

Nelson, J. A. (1996). The masculine mindset of economic analysis. *The Chronicle of Higher Education, 42,* B3.

Nepstad, D. C., Veríssimo, A., Alencar, A. et al (1999). Large-scale impoverishment of Amazonian forests by logging and fire. *Nature, 398,* 505–508.

Neumann, J., and Flohn, H. (1987). Great historical events that were significantly affected by the weather, part 8: Germany's war on the Soviet Union, 1941–1945. *Bulletin of the American Meteorological Society, 68,* 620–30.

Ni, J. (2001). Carbon storage in terrestrial ecosystems of China: Estimates at different spatial resolutions and their responses to climate change. *Climatic Change, 49,* 339–58.

Nicholls, N. (1998). William Stanley Jevons and the climate of Australia. *Australian Meteorological Magazine, 47,* 285–93.

Nirenberg, M. W. (1963). The genetic code: II. *Scientific American, 208* (March), 80–94.

Nordhaus, W. D. (1994). *Managing the global commons: The economics of climate change.* Cambridge, MA: MIT Press.

Nordhaus, W. D. (1994a). Expert opinion on climatic change. *American Scientist, 82,* 45–51.

Norgaard, R. B. (1990). Economic indicators of resource scarcity: a critical essay. *Journal of Environmental Economics and Management, 19,* 19–25.

Oppenheimer, M., and Alley, R. B. (2004). The west Antarctic ice sheet and long-term climate policy. *Climatic Change, 64,* 1–10.

Oreskes, N. (2000). Why predict?: Historical perspectives on prediction in the earth sciences. In D. Sarewitz, R. Pielke, Jr., and R. Byerly, Jr. (eds.), *Prediction: Decision-making and the future of nature.* Washington, DC: Island Press.

Oreskes, N., Shrader-Frechette, K., and Belitz, K. (1994). Verification, validation, and confirmation of numerical models in the earth sciences. *Science, 263,* 641–46.

Organisation for Economic Co-operation and Development. (2005). Economic Outlook No. 78, November 2005.

Orléan, A. (2001). Humility in economics. *Post-autistic economics newsletter, 5.* [http://www.paecon.net/PAEtexts/Orlean1.htm]

Ormerod, P. (1997). *The death of economics.* New York: Wiley.

Ormerod, P. (2000). *Butterfly economics.* New York: Basic Books.

Orrell, D. (2006). Post-Pythagorean Economics. *Adbusters, 67.*

Orrell, D. (2001). Modelling nonlinear dynamical systems: Chaos, error, and uncertainty. Ph.D. thesis, Oxford University.

Orrell, D. (2002). Role of the metric in forecast error growth: How chaotic is the weather? *Tellus, 54A,* 350–62.

Orrell, D. (2003). Model error and predictability over different timescales in the Lorenz '96 systems. *Journal of Atmospheric Science, 60,* 2219–28.

Orrell, D. (2005a). Ensemble forecasting in a system with model error. *Journal Atmospheric Science, 62,* 1652–59.

Orrell, D. (2005b). Filtering chaos: A technique for estimating observational and dynamical noise in nonlinear systems. *International Journal of Bifurcation and Chaos, 15,* 99–107.

Orrell, D. (2005c). Estimating error growth and shadow behavior in nonlinear dynamical systems. *International Journal of Bifurcation and Chaos, 15,* 3265–3280.

Orrell, D. (2006). Post-Pythagorean Economics. *Adbusters, 67.*

Orrell, D., Smith, L., Barkmeijer, J., and Palmer, T. (2001). Model error in weather forecasting. *Nonlinear Processes in Geophysics, 8,* 357–71.

Orrell, D., and Bolouri, H. 2004. Control of internal and external noise in genetic regulatory networks. *Journal of Theoretical Biology, 230,* 301–12.

Orrell, D., Ramsey, S., Marelli, M., Smith, J. J., Petersen, T. W., de Atauri, P., Aitchison, J. D., and Bolouri, H. (2005). Feedback control of stochastic noise in the yeast galactose utilization pathway, *Physica D, 217,* 64–76.

Orrell, D., Ramsey, S., de Atauri, P., and Bolouri, H. (2005a). A method to estimate stochastic noise in large genetic regulatory networks. *Bioinformatics, 21,* 208–17.

Osterholm, M. T. (2005). Preparing for the next pandemic. *Foreign Affairs, 84* (July/August).

Ovid. *Metamorphoses: Book the first.* Translated by Sir Samuel Garth, John Dryden, et al. <http://classics.mit.edu/Ovid/metam.1.first.html>.

Parmesan, C., and Galbraith, H. (2004). *Observed impacts of global climate change in the U.S.* Pew Centre on Global Climate Change. <http://www.pewclimate.org>.

Pearson, K. (1905). The problem of the random walk. *Nature, 72,* 294.

Peterson, I. (1993). *Newton's clock.* New York: W. H. Freeman.

Plato. (1961). *Collected dialogues.* Princeton, NJ: Princeton University Press.

Plender, J. (2005). Black clouds begin to gather where hedge funds tread. *Financial Times,* May 14, 2005.

Popper, K. R. (1957). *The poverty of historicism.* Boston: Beacon Press.

Population Reference Bureau. (2000). *Human population: Fundamentals of growth and change.* Washington, DC: Population Reference Bureau.

Powell, B. P. (1998). *Classical mythology.* Upper Saddle River, NJ: Prentice Hall.

Preston, R. (2002). *The demon in the freezer.* New York: Random House.

Rabier, F., Klinker, E., Courtier, P., and Hollingsworth, A. (1996). Sensitivity of forecast errors to initial conditions. *QJR Meteorological Society, 122,* 121–50.

Ramsey, S., Orrell, D., and Bolouri, H. (2005). Dizzy: Stochastic simulation of large-scale genetic regulatory networks. *Journal of Bioinformatics and Computational Biology, 3 (2),* 1–21.

Ramsey, S., Smith, J. J., Orrell, D., Marelli, M., Petersen, T. W., de Atauri, P., Bolouri, H. and Aitchison, J. D. (2006). Dual feedback loops in GAL regulon suppress cellular heterogeneity in yeast. *Nature Genetics, 38.*

Ravensbergen, J. (2004). Health pros prepare tools to confront flu epidemic. *Edmonton Journal,* December 6, 2004.

Ray, P. H., and Anderson, S. R. (2000). *The cultural creatives: How 50 million people are changing the world.* New York: Harmony Books.

Rayner, S. (2000). Prediction and other approaches to climate change policy. In D. Sarewitz, R. Pielke, Jr., and R. Byerly, Jr. (eds.), *Prediction: Decision-making and the future of nature.* Washington, DC: Island Press.

Reagan, N., with Novak, W. (1989). *My turn: The memoirs of Nancy Reagan.* New York: Random House.

Rees, M. (2003). *Our final hour.* New York: Basic Books.

Richardson, D. (1997). The relative effect of model and analysis differences on ECMWF and UKMO operational forecasts. In T. Palmer (ed.), *Predictability.* Reading, U.K.: European Centre for Medium-Range Weather Forecasting.

Richardson, L. F. (1922). *Weather prediction by numerical process.* Cambridge: Cambridge University Press.

Ridley, B. K. (2001). *On science.* London: Routledge.

Ridley, M. (1999). *Genome.* New York: HarperCollins.

Ridley, M. (2003). *Nature via nurture.* New York: HarperCollins.

Roach, J. (2001). Delphic oracle's lips may have been loosened by gas vapors. *National Geographic News,* August 14, 2001.

Roberts, H. (1959). Stock-market "patterns" and financial analysis: Methodological suggestions. *Journal of Finance, 14,* 1–10.

Roebber, P. J., Schultz, D. M., Colle, B. A., and Stensrud, D. J. (2004). Towards improved prediction: High-resolution and ensemble modeling systems in operations. *Weather Forecasting, 19,* 936–49.

Röhl, E. (1950). Los Diluvios en las Montañas de la Cordillera de la Costa. *Boletín de la Academia de Ciencias Físicas, Matemáticas y Naturales, Venezuela, 38,* 1–28.

Rosenthal, J. S. (2005). *Struck by lightning.* Toronto: HarperCollins.

Rotmans, J., and Dowlatabadi, H. (1998). Integrated assessment modeling. In S. Rayner and E. L. Malone (eds.), *Human choice and climate change, vol. 3: Tools for policy analysis.* Columbus, OH: Battelle Press.

Roulston, M. S., Kaplan, D. T., Hardenberg, J., and Smith, L. A. (2003). Using medium-range weather forecasts to improve the value of wind energy production. *Renewable Energy, 28,* 585–602.

Ruddiman, W. F. (2005). How did humans first alter global climate? *Scientific American* (March) 2005.

Sagan, C., and Turco, R. (1990). *A path where no man thought: Nuclear winter and the end of the arms race.* New York: Random House.

Saul, J. R. (1992). *Voltaire's bastards: The dictatorship of reason in the West.* Toronto: Penguin.

Saunders P. T., Koeslag J. H., Wessels J. A. (1998). Integral rein control in physiology. *Journal of Theoretical Biology, 194,* 163–73.

Schneider, S. H. (1989). *Global warming.* San Francisco: Sierra Club Books.

Schneider, S. H. (2002). Can we estimate the likelihood of climatic changes at 2100? *Climatic Change, 52,* 441–51.

Schneider, S. H., and Root, T. L. (2001). Climate change: Overview and implications for wildlife. In S. H. Schneider and T. L. Root (eds.), *Wildlife responses to climate change: North American case studies.* Washington, DC: Island Press.

Schnur, R. (2002). The investment forecast. *Nature, 415,* 483–84.

Schrödinger, E. (1944). *What is life?* Cambridge: Cambridge University Press.

Seaton, R. A. (2004). *Western philosophy.* Unpublished thesis.

Shamas, L. (2003). Cassandra ignored/Apollo avenged: The complex of disbelief in 2003. Headlinemuse.com. <http://www.headlinemuse.com/Culture/cassandra.htm>.

Shapiro, J. (1999). Genome system architecture and natural genetic engineering in evolution. In Lynn Helena Caporale (ed.), Molecular strategies in biological evolution, *Annals of the New York Academy of Sciences, 870.*

Sherden, W. A. (1998). *The fortune sellers: The big business of buying and selling predictions.* New York: John Wiley.

Shiva, V. (2002). *Water wars: Privatization, pollution, and profit.* Cambridge, MA: South End Press.

Simmons, A., Mureau, R., and Petroliagis, T. (1995). Error growth and estimates of predictability from the ECMWF forecasting system. *QJR Meteorological Society, 121,* 1739–71.

Simons, K. (1997). Model error. *New England Economic Review,* November 17–28, 1997.

Singer, D. (2005). Private communication. [e-mail, August 2, 2005.]

Skinner, B. F. (1973). *Beyond freedom and dignity.* Harmondsworth, U.K.: Penguin.

Slonim, D. K., Tamayo, P., Mesirov, J. P., Golub, T. R., and Lander, E. S. (2000). Class prediction and discovery using gene expression data. Proceedings of the fourth annual International Conference on Computational Molecular Biology, Tokyo, Japan, 263–72.

Smith, A. (1776). *An inquiry into the nature and causes of the wealth of nations.* London: Methuen. (Reprinted in 1999 as *The Wealth of Nations: Books I–III* by Penguin, London, U.K.)

Smith, L. A. (1996). Accountability in ensemble prediction. In T. Palmer (ed.), *Predictability.* Reading, U.K.: European Centre for Medium-Range Weather Forecasting.

Smith, L. A. (1997). The maintenance of uncertainty. In G. C. Castagnoli and A. Provenzale (eds.), *Past and present variability in the solar-terrestrial system: Measurement, data analysis and theoretical models,* vol. 133 of the International School of Physics, Il Nuovo Cimento, Bologna, Italy.

Smith, L. A. (2000). Disentangling uncertainty and error: On the predictability of nonlinear systems. In A. I. Mees (ed.), *Nonlinear dynamics and statistics.* Boston: Birkhauser.

Smith, L. A., Ziehmann, C., and Fraedrich, K. (1999). Uncertainty dynamics and predictability in chaotic systems. *QJR Meteorological Society, 125,* 2855–86.

Soros, G. (2000). *Open society: Reforming global capitalism.* New York: Public Affairs.

Stainforth, D. A., Aina, T., Christensen, C., Collins, M., Frame, D. J., Kettleborough, J. A., Knight, S., Martin, A., Murphy, J., Piani, C., Sexton, D., Smith, L. A., Spicer, R. A., Thorpe, A. J., and Allen, M. R. (2005). Uncertainty in predictions of the climate response to rising levels of greenhouse gases. *Nature, 433,* 403–406.

Stewart, I. (1989). *Does God play dice: The new mathematics of chaos.* Cambridge: Blackwell.

Stoto, M. (1983). The accuracy of population projections. *Journal of the American Statistical Association, 78,* 13–20.

Strohmeir, J., and Westbrook, P. (1999). *Divine harmony: The life and teachings of Pythagoras.* Berkeley, CA: Berkeley Hills Books.

Sunnafrank, M., and Ramirez., A. (2004). At first sight: Persistent relational effects of get-acquainted conversations. *Journal of Social and Personal Relationships, 21,* 361–79.

Suzuki, D. (2005). The beauty of wind farms. *New Scientist,* April 16, 2005.

Taverne, D. (2005). *The march of unreason.* Oxford: Oxford University Press.

Tetlock, P. E. (2005). *Expert Political Judgment: How Good Is It? How Can We Know?* Princeton: Princeton University Press.

Thelen, E., and Smith, L. B. (1994). *A dynamic systems approach to the development of cognition and action.* Cambridge, MA: MIT Press.

Thucydides (1910). *The Peloponnesian War.* London: J. M. Dent.

Tjio, J. H., and Levan, A. (1956). The chromosome number of man. *Hereditas, 42,* 1.

Tol, R. S. J., and Dowlatabadi, H. (2002). Vector-borne diseases, development, and climate change. *Integrated Environmental Assessment, 2,* 173–81.

Toth, Z., Kalnay, E., Tracton, S., Wobus, S., and Irwin, J. (1996). A synoptic evaluation of the NCEP ensemble. In T. Palmer (ed.), *Proceedings of the 5th Workshop on Meteorological Operating Systems.* Reading, U.K.: European Centre for Medium-Range Weather Forecasting.

Trenberth, K. (2005). Uncertainty in hurricanes and global warming. *Science, 308,* 1753–54.

Tumpey, T. M., et al (2005). Characterization of the reconstructed 1918 Spanish influenza pandemic virus. *Science, 310,* 77–80.

Union of Concerned Scientists (1993). *World scientists' warning to humanity.* Cambridge, MA: Union of Concerned Scientists.

Uttal, T., et al (2002). Surface heat budget of the Arctic Ocean. *Bulletin of the American Meteorological Society, 83,* 255–75.

Van der Waerden, B. L. (1974). *Science awakening 2: The birth of astronomy.* New York: Oxford University Press.

Van Kampen, K. (2005). *The golden cell.* Toronto: HarperCollins.

Vedantam, S. (2004). Warning system not in place. *Washington Post,* December 27, 2004.

Verma, M., Bhat, P. J. and Venkatesh, K.V. (2003). Quantitative analysis of gal genetic switch of *saccharomyces cerevisiae* reveals that nuclearcytoplasmic shuttling of gal80p results in a highly sensitive response to galactose. *Journal of Biological Chemistry, 278,* 48764–48769.

Vernadsky, V. I. (1929). *La biosphere.* Paris: Felix Alcan.

Vine, B. (2005). *The minotaur.* London: Viking.

Walker, G. T. (1918). Correlation in seasonal variations of weather. *Quarterly Journal of the Royal Meteorological Society, 44,* 223–24.

Wallace, P. (2000). Just wait for the gold rush to end. *New Statesman,* February 21, 2000.

Watson, A. J., and Lovelock, J. E. (1983). Biological homeostasis of the global environment: The parable of Daisyworld. *Tellus, 35B,* 284.

Watson, J. D. (1968). *The double helix.* London: Readers Union.

Watson, J. D., and Crick, F. H. C. (1953). Molecular structure of nucleic acids: A structure for deoxyribose nucleic acid. *Nature, 171,* 737.

Watson, W. (2005). Bears bash Bush—over the events of 2080. *National Post,* September 29, 2005.

Watts, D. J. (2003). *Six degrees: The science of a connected age.* New York: W. W. Norton.

Webster, P. J., Holland, G. J., Curry, J. A., and Chang, H.-R. (2005). Changes in tropical cyclone number, duration, and intensity in a warming environment. *Science, 309,* 1844–46.

Wertheim, M. (1995). *Pythagoras' trousers: God, physics, and the gender wars.* New York: W. W. Norton.

Weston, A. D., and Hood, L. (2004). Systems biology, proteomics, and the future of health care: Toward predictive, preventative, and personalized medicine. *Journal of Proteome Research, 3,* 179–96.

White, A. D. (1896). *A history of the warfare between science and theology in Christendom.* New York: D. Appleton-Century Company.

Whitfield, J. (2003). Too hot to handle. *Nature, 425,* 338–39.

Wichman, H. A., Badgett, M. R., Scott, L. A., Boulianne, C. M., and Bull, J. J. (1999). Different trajectories of parallel evolution during viral adaptation. *Science, 285,* 422–24.

Wieczorek, G. F., Larsen, M. C., Eaton, L. S., Morgan, B. A., and Blair, J. L. (2001). Debris-flow and flooding hazards associated with the December 1999 storm in coastal Venezuela and strategies for mitigation. U.S. Geological Survey Open File, Report 01–144.

Williams, J. B. (1938). *The theory of investment value.* Cambridge, MA: Harvard University Press.

Willich, S. N., Wegscheider, K., Stallmann, M., and Keil, T. (2006). Noise burden and the risk of myocardial infarction. *European Heart Journal, 27,* 276–282.

Wohlforth, C. (2004). *The whale and the supercomputer: On the northern edge of climate change.* New York: North Point Press.

Wolfram, S. (2002). *A new kind of science.* Champaign, IL: Wolfram Media.

Wood, M. (2003). *The road to Delphi.* New York: Farrar, Straus, and Giroux.

Worldwatch Institute (1998). *State of the world.* Edited by Lester Brown et al. New York: W. W. Norton.

Woudenberg, F. (1991). An evaluation of Delphi. *Technological Forecasting and Social Change* (September), 131.

Wright, R. (2004). *A short history of progress.* Toronto: Anansi.

Yamamoto, K. R. (1985). Steroid receptor regulated transcription of specific genes and gene networks. *Annual Review of Genetics, 19,* 209–52.

Yorke J., and Li, T. Y. (1975). Period three implies chaos. *American Mathematical Monthly, 82,* 985–92.

Zhang, J. V. et al. (2005). Obestatin, a peptide encoded by the ghrelin gene, opposes ghrelin's effects on food intake. *Science, 310*, 996–999.

Ziegler, P. (1965). *The black death.* New York: Harper and Row.

Zipf, G. K. (1949). *Human behavior and the principle of least effort.* Cambridge, MA: Addison-Wesley.

Zöllner, F., and Nathan, J. (2003). *Leonardo da Vinci: The complete paintings and drawings.* Cologne: Taschen.

Acknowledgements

For helping this book to exist, and make sense, thanks to my agent, Robert Lecker; my editor, Jim Gifford; copy-editor Janice Weaver; managing editor Noelle Zitzer; and the team at HarperCollins Canada.

For generously sharing their time and knowledge, I thank (without implying their agreement with the contents) Pedro de Atauri, Hamid Bolouri, Jack von Borstel, Beatriz Leon, Robert Matthews, Joe McCauley, Wendy Orrell, Stephen Ramsey, and Lenny Smith.

Thanks also to the European Centre for Medium-Range Weather Forecasts in the United Kingdom and the Institute for Systems Biology in Seattle, Washington.

Index

Made in the USA
Lexington, KY
02 April 2013